Practical Math in Context

Book 3

On the Job

SADDLEBACK PUBLISHING • INC.

Development: Frishman Co
Authors: Lucia McKay, Ph.D. and Maggie Guscott
Design and Production: The Format Group, LLC
Cover Art: IQ Design

SADDLEBACK PUBLISHING, INC.
Three Watson
Irvine, CA 92618-2767
Web site: www.sdlback.com

ISBN 1-56254-764-X

Printed in the United States of America

10 09 08 07 06 9 8 7 6 5 4 3 2 1

Table of Contents

Unit 5: Getting to the Job

Unit 6: Getting Paid

Unit 7: Teamwork

Unit 8: Unemployment

To the Student

Welcome to *On the Job!* This is Book 3 of the *Practical Math in Context* series.

The goal of this book and the other books in this series is to build your confidence and practical math skills. You will use these math skills in everyday situations throughout your life.

You solve problems and make mathematical decisions every day. You compare products and make choices about what to buy. You work to earn money. You decide what to spend and how much to save.

Practical Math in Context gives you strategies to solve everyday math problems in a variety of ways. It strengthens your skills and gives you practice with many different math topics. Each of the six books presents topics you are likely to encounter in everyday life. Each book includes problems that involve estimation, equations, mental math, calculators, and critical thinking. Each book includes additional topic-specific skills such as graphing, averages, statistics, ratios, and measurement.

Each unit begins with a preview lesson, which models and explains the types of problems you will encounter in the unit. Then there are five lessons, at least one of which is usually a game. Each unit ends with a review of the unit concepts. There are illustrations and graphic art throughout.

Here is a list of the titles of the other books in the *Practical Math in Context* series:

Book 1: Everyday Life
Book 2: Home & School
Book 4: Budgeting & Banking
Book 5: Smart Shopping
Book 6: Sports, Hobbies, & Recreation

With review and practice, you will build your math skills and learn to approach everyday mathematical situations with confidence! *Practical Math in Context* will help you become a successful problem solver!

Unit 1 Career Options

Preview

How You Will Use This Unit

As you think about career options, you will consider many different things. Salary and take-home pay are just two examples. You will probably also consider types of work and benefits that come with the job. You may also think about whether to work indoors or outdoors and how you might work with people or products. As you compare options and make choices, you will often use math. The math skills you use include mental math and estimation, basic operations and equations, statistics and probability, and measurement of circles.

What You Will Do in This Unit

In this unit, math steps demonstrate how to solve problems. These steps can help you answer questions such as these:

You run an express delivery service. You pick up a package in the northeast quadrant of the city and deliver it to an address in the southeast quadrant. In what general direction do you head?

You work at the local bookstore. You earn $12,500 a year. There is a job opening at the big bookstore downtown. The salary is $14,000. What is the percent increase in pay?

You are a dental assistant. You prefer to work only in the afternoons. Each week, the schedule changes. What is the probability of working only in the afternoons three weeks in a row?

You work for a local construction company. During the week, the company pays you $12 an hour. On Saturdays, you do charity work at half pay. How much do you earn on Saturdays?

What You Can Learn from This Unit

When you complete this unit, you will have used mathematics to solve problems related to career options. These problems are similar to those that may actually occur in your daily life.

Lesson 1 Outdoor Careers

Example Enrique is a paramedic. He flies in a helicopter that does emergency pickups of hurt or sick people. Today, he heads in a direction 245° from the base. Mark that heading on this diagram. In which direction is Enrique heading (north, northeast, east, southeast, south, southwest, west, or northwest)?

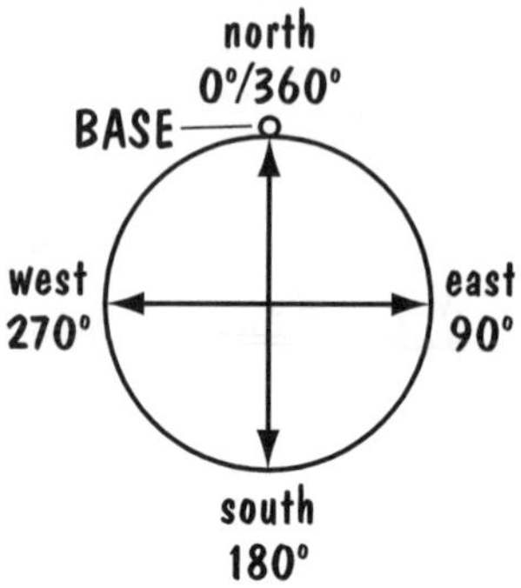

Solve

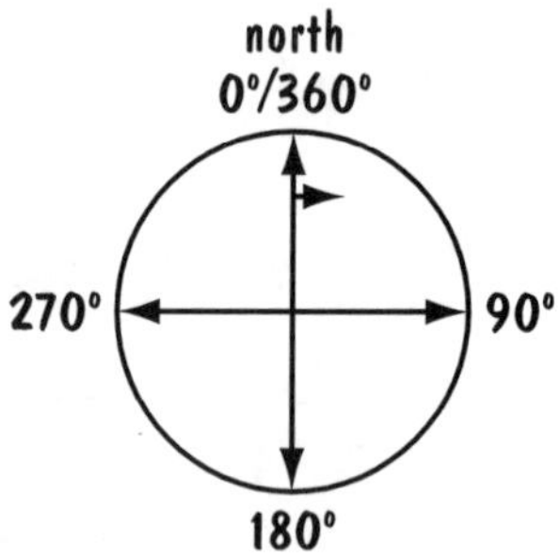

Step 1: Headings are always measured in a clockwise direction starting from north at 0°. Draw a small arrow to show where to start measuring Enrique's direction.

Step 2: Now, plot Enrique's heading. You can use a protractor to draw the 245° angle from 0° north. OR you can subtract the angle from 360° north and plot that angle in the negative (counter-clockwise) direction.

$360° - 245° = 115°$

Answer the Question

Step 3: Enrique is heading toward the southwest.

✏ Now try these problems.

1. Bette runs an express delivery service around the city. She picks up a package in the northeast quadrant of the city. She delivers it to an address in the southeast quadrant. In what general direction should she head?

   Answer: Generally, Bette heads toward the ___________.

2. Margo conducts walking tours of the island. Before a tour, she checks the heat index on the following chart. When the heat index is above 100, she does indoor tours of the museums. Otherwise she does outdoor tours. On the chart, the heat index is in the cell where a row and column meet.

a. Yesterday morning, the air temperature was 85° and the relative humidity was 80%. Circle the heat index.

Air Temp (°F)	Relative Humidity (%) 45	50	55	60	65	70	75	80	85	90	95	100
95°	104	107	110	114	119	124	130	136				
90°	95	96	98	100	102	106	109	113	117	122		
85°	87	88	89	90	91	93	95	97	99	102	105	108
80°	80	81	81	82	83	85	86	86	87	88	89	91

b. This morning, the relative humidity is 65%. According to this chart, how high can the air temperature be for Margo still to do outdoor tours?

A 80° **B** 85° **C** 90° **D** 91°

3. Kael earns $6,500 over the four-month ski season as a ski instructor. During the rest of the year he earns $1,875 a month as a sports-club coach. What is his annual income before taxes?

Answer: ___________

4. Terry is head groundskeeper for the city's park system. He works 8 hours a day and earns $12 an hour. He works about 220 days a year. On the job, he walks an average of 11 miles a day. About how many miles does he walk in one year? What is the mean (average) rate of pay per mile? Draw a line to join the correct pair of answers.

132 miles	$8.73 per mile
2,420 miles	$96 per mile
2,640 miles	$20 per day
4,015 miles	$27.50 per day

☆ *Challenge Problem*

You may want to talk this one over with a partner.

The salary of one leading baseball player was $120,000 in 1997. In 2002 it was $1,250,000. He says that each year his salary *increase* is twice the increase of the year before. Find his salary in 2003. **Answer:** ___________

Lesson 2

Indoor Careers

Example Marnie is the librarian in the downtown library. She earns \$24,690 a year. There is an opening for a librarian at the local high school. The salary is \$27,400. Marnie interviews and gets the job. What is the percent increase in her salary?

Solve

Step 1: Find the *amount* of increase from Marnie's old salary to her new salary.

\$27,400 – \$24,690 = \$2,710

Step 2: Now, write an expression for percent increase.

$$\text{Percent increase} = \frac{\text{Amount of increase}}{\text{Original salary}} \times 100\%$$

Step 3: Substitute the amount of increase and original salary.

$$\text{Percent increase} = \frac{\$2{,}710}{\$24{,}690} \times 100\% = 10.98\%$$

Answer the Question

Step 4: The percent increase in Marnie's salary is 10.98%.

Now try these problems.

1. Lamar is a chef at a hotel in the suburbs. He earns \$20,200 a year. There is an opening for head chef at a hotel downtown. The salary is \$29,800. Lamar interviews and gets the job. What is the percent increase in his salary?

   Answer: The percent increase in Lamar's salary is ________%.

2. Charlie suggests that his legal assistant go to a paralegal seminar. She submits an expense request for this trip to the

seminar. Complete the entries for the blank lines on the request. How much will the trip cost?

EXPENSE REQUEST

Item	Price Detail	Quantity	Estimated Total
Airfare:	round trip		$250
Hotel:	$85 per day	3 days	$________
Meal allowance:	$9.50 per meal	9	$________
Mileage:	$0.28 cents a mile	50 miles	$________
Airport parking:	maximum of $25		$25
Telephone:	maximum of $25		$10
TOTAL			$________

3. Darien is a bank teller. She works from 8:45 a.m. to 4:15 p.m. five days a week. On Saturdays, she volunteers at the local animal shelter for 8 hours. How many hours does she work each week, including her volunteer work?

A 15.50 **B** 16 **C** 45.5 **D** 48

4. The state law says a minor cannot spend more than 40 hours a week going to school and working. Kyla goes to school between 7:45 a.m. and 1:15 p.m., Monday through Friday. She works in a doctor's office every afternoon, starting at 2:00 p.m. At what time does she leave each day in order to obey the state law? Mark that time on the clock face.

☆ *Challenge Problem*

You may want to talk this one over with a partner.

An e-mail ad says, "Work from home online and earn $2,500 per month part-time." What does "part-time" mean to you? How many work days does a "month" mean to you? Use your definitions and write an expression for the rate of pay per hour for this job. Suppose you worked twice as many hours to do this job. Would that mean that your income goes up or that your hourly rate of pay goes down? Explain.

____

____

____

Lesson 3 — Working with People

Example Corey is a dental assistant. He prefers to work *only* in the afternoons. Each week the office assigns him randomly to any one of their five different work schedules. What is the probability that he gets to work *only* in the afternoons three weeks in a row?

Work Schedules

Schedule	Start	Finish
A	6:00 a.m.	2:00 p.m.
B	7:30 a.m.	3:30 p.m.
C	9:00 a.m.	5:00 p.m.
D	11:00 a.m.	7:00 p.m.
E	1:00 p.m.	9:00 p.m.

Solve

Step 1: Underline the schedule(s) that Corey prefers.

<u>E 1 p.m. 9 p.m.</u>

Step 2: Write the probability that he gets this schedule the first week.

$\frac{1}{5}$ Probability = $\frac{\text{favorable outcome}}{\text{total possible outcomes}}$

Step 3: Write the probability that he gets this schedule three weeks in a row.

$\frac{1}{5} \times \frac{1}{5} \times \frac{1}{5}$

$\frac{1}{125}$

Answer the Question

Step 4: The probability that Corey gets to work *only* in the afternoons for three weeks in a row is $\frac{1}{125}$.

Now try these problems.

1. Myrna is one of four night-shift customer service engineers. Each takes a two-hour shift between 11 p.m. and 7 a.m. at the online help desk. They toss a number cube each day to determine which shift each will take. What is the probability that Myrna will get the same shift four days in a row?

Answer: The probability is ________.

2. Greg now earns $12 an hour as a lifeguard. He works 30 hours/week, 39 weeks/year. He plans to complete a teacher's certificate soon. He can earn $30,000 as a first-year teacher. He calculates the percent increase from his present salary to a teacher's salary. Fill in the blanks to find the percent increase.

Difference between teacher's salary and lifeguard salary = $________

Percent increase in income = [$_______ ÷ $_______] × 100% = _____%

3. Rachael is a sports reporter. When the wind chill falls below –24°F, games are postponed.

a. Circle all the cells having a wind chill that means a game is postponed.

Wind speed (mph)	Actual thermometer reading (°F) 40	30	20	10	0	–10	–20	–30	–40
5	37	27	16	6	–5	–15	–26	–36	–47
10	28	16	4	–9	–21	–33	–46	–58	–70
15	22	9	–5	–18	–36	–45	–58	–72	–85
20	18	4	–10	–25	–39	–53	–67	–82	–96

You will find the wind chill (for a given temperature and wind speed) in the cell where a row and column meet.

b. Today the thermometer reading is 10°F, and the wind speed is 20 mph. Is today's game held or postponed? Check the correct box.

Answer: Held ☐ Postponed ☐

4. Faith earns $100 an hour playing the piano for a dance school. She also works part-time as a physical therapist and earns $18,000 a year. How many hours of playing the piano does it take for her to double her total yearly income?

A 20 **B** 180 **C** 200 **D** 360

☆ *Challenge Problem*

You may want to talk this one over with a partner.

Rick and Gabe, compete for the greatest average (mean) sales per day in one week. Gabe still has to report his sales for Friday. How much does Gabe have to sell on Friday to beat Rick? Explain your answer.

	Monday	Tuesday	Wednesday	Thursday	Friday
Rick	$1,208	$734	$525	$ 904	$450
Gabe	$ 384	$980	$975	$1,050	

____

____

Lesson 4

Career Tracks

A Board Game (for Two or More Players)

The goal of this game is to place five counters in a track. On the game board, the track can be a radius, an arc of a circle, or a spiral. (Examples of each are shown on the next page.) First, you place a counter at one intersection. Then you place your next counter at another intersection.

Materials

Game board (on the next page), twelve counters in each of several colors

Directions

1. Choose a counter color for each player. Sit with players around the game board.
2. Player 1 places one counter at a line-circle intersection.
3. Player 2 then takes a turn and places a counter at another line-circle intersection.
4. Players take turns placing their counters. You can try to block the track you think another player is trying to build. You block by placing one of your counters at a line-circle intersection. If you cannot develop this counter into a track, then it is a lost counter, or a lost career move.
5. The winner is the first player to have five counters making a track. Discuss the career track the winner built.

Before you play the game, try these warm-up problems.

1. Three players are playing *Career Tracks.* Each player has four counters on the game board. Only one player has been blocked one time. In how many turns can there be a winner? Explain your answer.

   Answer: ______________________________

2. Taylor starts at the entry level sports coach position. She plans a spiral career track. In five moves where could she be?

   A firefighter　　**C** car mechanic

   B legal assistant　　**D** groundskeeper

Game Board for *Career Tracks*

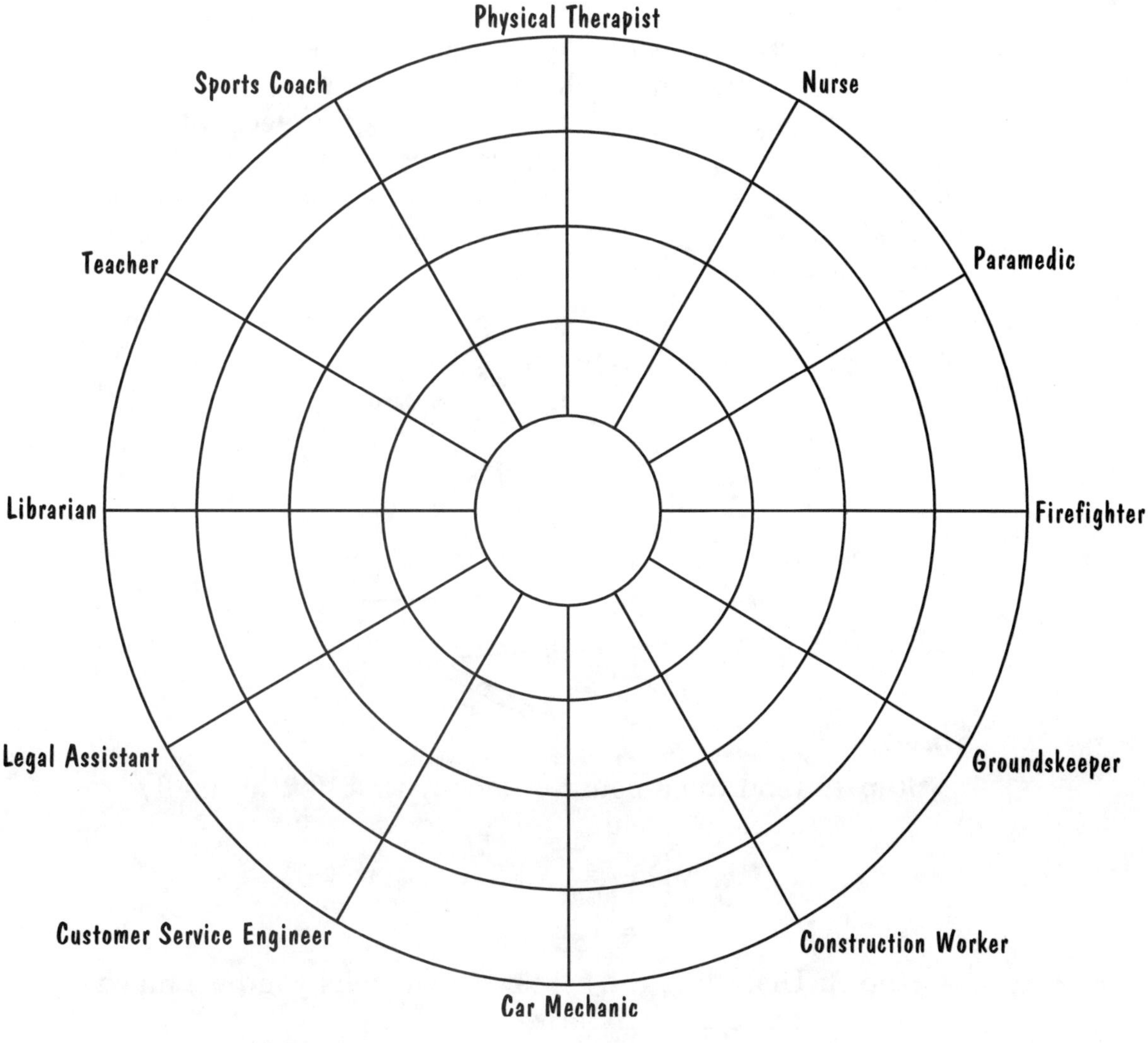

Legend:

The innermost circle represents entry-level positions.

The third circle represents supervisory positions.

The outermost level represents upper-management positions.

The levels in between are transitional positions.

Radius: straight line career from the ground up to the top of the hierarchy.

Arc of a circle: Transition to other associated careers at the same level.

Spiral: Progressive advancement up the hierarchy, and transition through other associated careers.

Example Tracks:

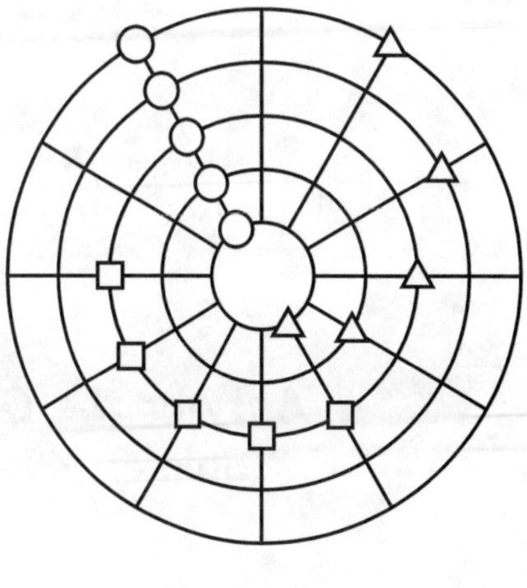

Lesson 5

Products, Routines, & Pay

Example Jeff works for his dad's construction company. During the week, the company assigns him to jobs where the customer is charged \$30 an hour for his work. The company pays Jeff 40% of that. On Saturdays, they do charity projects at half the normal cost, and Jeff earns half his normal pay. The company pays a \$25 bonus to any employee who participates in a Saturday project. How much does Jeff earn for one eight-hour Saturday workday?

Solve

Step 1: Find the amount of money that Jeff normally earns per hour.

$40\% \text{ of } \$30 = \frac{40}{100} \times \$30$

$= \$12$

Step 2: Underline the sentence that tells you how much Jeff earns on Saturdays.

<u>On Saturdays ... Jeff earns half his normal pay.</u>

Step 3: Write the same sentence replacing "half his normal pay" with the actual amount.

Jeff earns half of \$12.

So Jeff earns $\frac{\$12}{2} = \$6$ an hour.

Step 4: Now, calculate the amount of money Jeff earns in eight hours on Saturday. (Remember the \$25 bonus for participation.)

$\$6 \times 8 + \$25 = \$73$

Answer the Question

Step 5: Jeff earns \$73 on a Saturday.

✏ Now try these problems.

1. Ramone works for his uncle's marine engine company. During the week, the company charges him out at $40 an hour and pays him 50% of that. On Saturdays, they do charity projects at one quarter of the normal price. For these projects, the company pays Ramone half his normal pay. The company pays a $30 bonus to any employee who participates in a Saturday project. How much does Ramone earn for one six-hour Saturday workday? Fill in the blanks.

   Answer: Ramone earns $______ × ______ + $______ = $______ for a six-hour Saturday workday.

2. Heather works in a circular showroom for wallpaper and tiles. Aisles divide the showroom like the spokes of a wheel. Each spoke aisle is 25 feet long. A spoke aisle goes from the center to the outer corridor that circles the entire showroom. In one eight-hour workday, Heather estimates that she walks around the outer corridor at least 17 times, and walks the length of six spoke aisles. What is the approximate distance that she walks in one day? ($C = 2\pi r$)

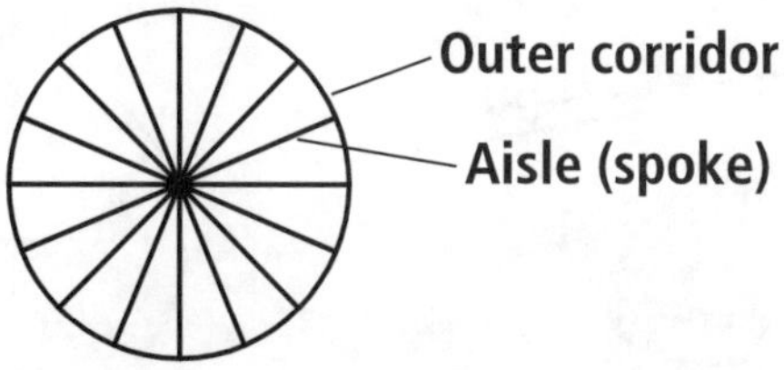

   A 150 feet

   B 850 feet

   C 2,669 feet

   D 2,819 feet

3. Kim is a digital photo specialist. *PhotoSharp* retains her for $3,000 a month to promote and sell digital cameras. The average (mean) sales price of the digital cameras they sell is $154. The average (mean) cost that *PhotoSharp* buys the cameras for is $45. How many cameras must Kim sell per month for the profits to equal or exceed her salary? Use the symbols, +, –, ×, ÷, to fill in the blanks and complete the equation.

   Answer: $3,000 ____ ($154 ____ $45) ≈ ______

☆ *Challenge Problem*

You may want to talk this one over with a partner.

Roy runs a car service business. Each of his mechanics works an eight-hour day. They take a one-hour lunch break together that is not included in their eight hours. He keeps exactly four of his mechanics busy on the shop floor at any one time. They rotate so that one mechanic is always delivering cars. What is the minimum number of mechanics Roy can employ? Draw a diagram to explain your answer.

Answer: ________

Unit 1 Review

Review What You Learned

In this unit you have used mathematics to solve many problems. You have used mental math and estimation, practiced basic operations, and solved equations. You have also used statistics and probability, and measurement of circles.

These two pages give you a chance to review the mathematics you used and check your skills.

✔ Check Your Skills

1. Louis is a firefighter. He regularly takes the helicopter to check for fires or to help people in difficulty. Today, he heads out from the base in a direction of 290°. Mark that heading on the diagram. In which direction is he heading (north, northeast, east, southeast, south, southwest, west, or northwest)?

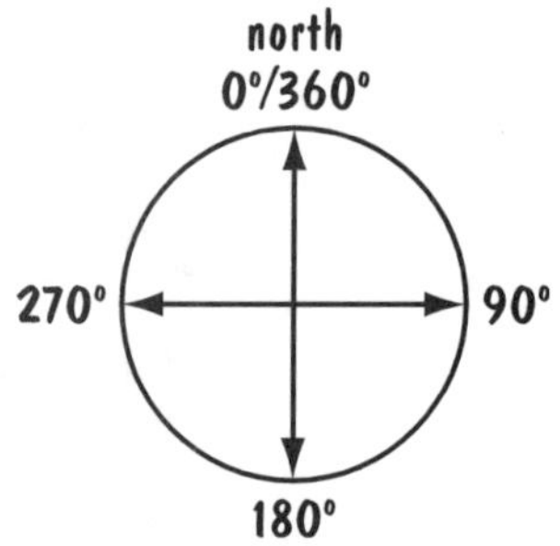

Answer: ________________.

If you need to review, return to lesson 1 (page 2).

2. Elton works at a specialty bookstore in town. He earns $12,300 a year. There is a job opening at the bookstore downtown. The salary is $14,500. He writes an expression to find out the percent increase in salary. Is his expression correct? If not, correct it, and compute the answer.

$$(\$14{,}500 - \$12{,}300) \times \$14{,}500 \div 100\%$$

Answer: ____ Yes ____ No

____

If you need to review, return to lesson 2 (page 4).

3. Courton is a physical therapist. He prefers a work schedule that allows him to leave by 3 p.m. at the latest. Each week the hospital randomly assigns him to one of their four different work

schedules. What is the probability that he gets to leave work by 3 p.m. three weeks in a row?

Work Schedules

Schedule	Start	Finish
A	7:00 a.m.	12:00 noon
B	10:00 a.m.	3:00 p.m.
C	1:00 p.m.	5:00 p.m.
D	3:00 p.m.	9:00 p.m.

Answer: ______

If you need to review, return to lesson 3 (page 6).

4. Beki works for her mother's public relations firm. During the week, the firm charges Beki out at $35 an hour, and pays her 60% of that. On weekends, they do charity projects at half the normal cost. Beki earns half her normal pay. The firm pays a $30 bonus to any employee who participates in a weekend project. How much does Beki earn for one ten-hour weekend project?

   A $105 **B** $135 **C** $175 **D** $205

If you need to review, return to lesson 5 (page 10).

Write Your Own Problem ✍

Choose a problem you liked from this unit. Write a similar problem using a situation and related facts from your own life. With a partner, share and solve these problems together. Discuss the mathematics and compare the steps you used. If you need to, rewrite or correct the problems. Write your edited problem and the answer here.

Unit 2 Providing Product or Service —

Preview

How You Will Use This Unit

Different industries offer many kinds of jobs. You might build products or provide a service. You might sell products, or fix things for people. You can rent vehicles or equipment to others. As you compare and make choices about jobs, you will often use math. The math skills you use include mental math and estimation, basic operations and equations, and statistics and probability. You will also use measurements on figures of different shapes with both standard and metric measurements.

What You Will Do in This Unit

In this unit, math steps demonstrate how to solve problems. These steps can help you answer questions such as these:

You buy tiles for the floor of your new cabana. The tiles are $\frac{1}{2}$-meter squares. The floor measures 8 meters by 6 meters. How many tiles do you need?

After school, you fix cars. You charge $7 per hour. An average job takes 5 hours. On Saturdays, you add a weekend charge of $20 more for the average job. What is your hourly rate on Saturdays?

You want to buy a refrigerator for the office. The refrigerator costs $499. You can pay for it outright, or you can make monthly payments. Which way of paying costs less, and by how much?

You are a sales person. You earn a base salary of $23,000 per year, plus a commission of 3% on all your sales. This year you earn a total of $35,000. What is the dollar value of your sales?

What You Can Learn from This Unit

When you complete this unit, you will have used mathematics to work problems related to building products, providing services, or renting vehicles or equipment. These problems are similar to those that may actually occur in your daily life.

Lesson 1

Building Things

Example Lucinda designs a rectangular cabana for her property. Her property is 21 meters by 27 meters. The floor space of the cabana is 8 meters by 6 meters. A building must be set back at least 3 meters from the street. A building must also be at least 2 meters from the property lines. Lucinda wants her cabana to be at least one meter from the main house. Where can she put it? Choose a spot and draw it to scale on the plan of her property.

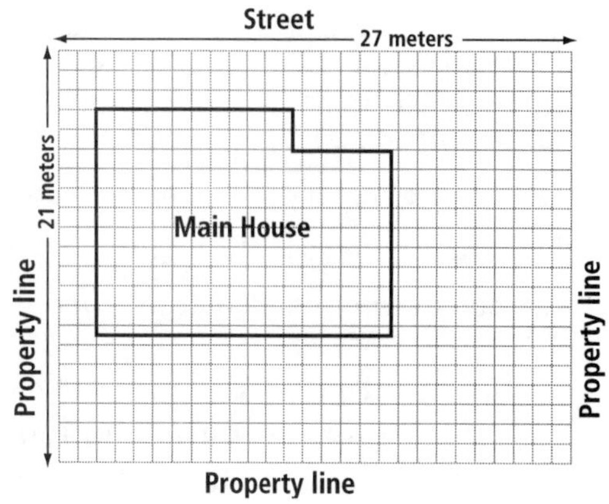

Solve

Step 1: Use these steps to show where she can build on the property. First, count 3 meters back from the street and draw that (set back) line.

Step 2: Next, count 2 meters in from each property line. Then draw each of these lines.

Step 3: Now, cut a piece of paper 8 meters by 6 meters (or 8 squares by 6 squares) to show Lucinda's cabana. See where the square will fit in the remaining area.

Step 4: Draw the cabana in on the plan of the property.

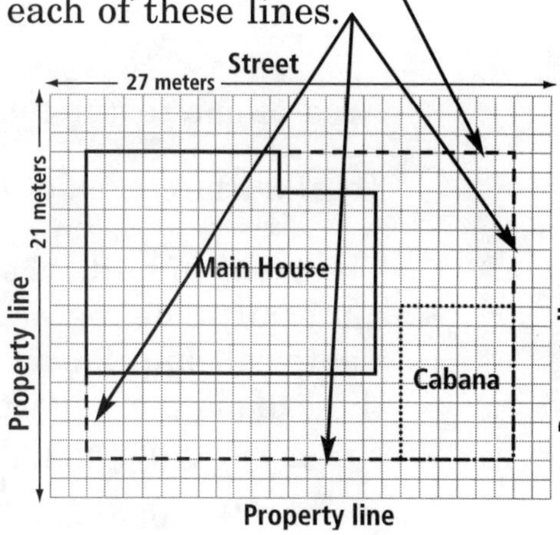

Answer the Question

Step 5: Lucinda can put her cabana anywhere in the strip to the right of the main house. This strip is 16 meters by 6 meters.

Now try these problems.

Use Lucinda's property plan.

1. Lucinda chooses square tiles for the floor of her new 8-meter by 6-meter cabana. The tiles are $\frac{1}{2}$ meter on each side. Write an expression for the number of tiles she needs.

Answer: Lucinda needs ____________________ = ______ tiles for her cabana floor.

2. The walls of Lucinda's cabana are 4 meters high. The front door is 2.3 meters tall and 1 meter wide. Two windows are each 1.5 meter high and 2 meters wide. The materials to build the walls cost $45 per square meter. Circle the line showing the correct dimensions and cost of the wall materials.

Wall Area (including door and windows)	Area of door	Area of (2) windows	Area requiring wall materials	Cost of materials
112 square meters	3.3 square meters	3 square meters	105.7 square meters	$475.60
112 square meters	2.3 square meters	6 square meters	103.7 square meters	$4,666.50
128 square meters	2.3 square meters	6 square meters	119.7 square meters	$5,386.50

3. The roof of the cabana is a gable roof. The gable roof is 1.5 meters high. The length of the slope is 3.4 meters. What is the total surface area (the four faces open to the sky) of the roof?

Area of a triangular end ($\frac{1}{2}$ base × height) = $\frac{1}{2}$ × ______ × ______ square meters.

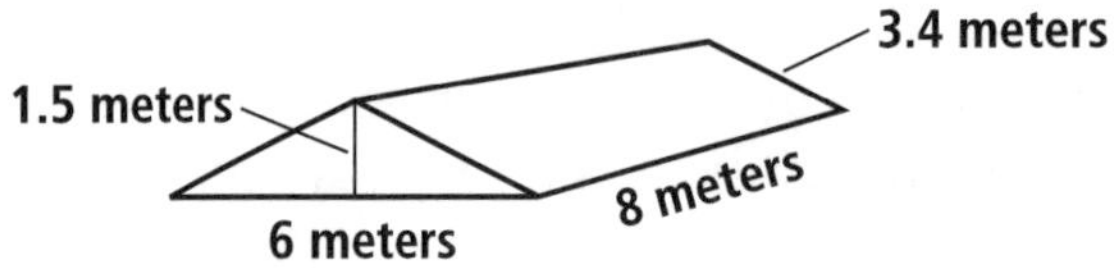

Area of one rectangular face = ______ × ______ square meters.

Total surface area of the roof = ______ square meters.

4. The labor to build the cabana costs 60% of the cost of the materials. The materials cost $9,200. What is the total cost (materials and labor) to build the cabana? What is the cost of the cabana per square meter of floor space?

   A $5,520; $306.67
   B $9,200; $525.71
   C $14,720; $306.67
   D $14,720; $525.71

☆ *Challenge Problem*

You may want to talk this one over with a partner.

You want to build a similar cabana to Lucinda's, but you want to spend about half of the money. What would you change to achieve this? Explain your answer.

____

____

____

Lesson 2

Providing a Service

Example Shane pays his mechanics \$25 per hour and charges customers \$65 per hour per mechanic. On the third Thursday of each month, he offers a special discount on a tune-up and oil change. The customer pays \$40 less for this job than on other days. (Shane still pays his mechanics \$25 per hour.) A tune-up and oil change takes one mechanic two hours. What is the change in the markup per hour on Thursdays for the work of Shane's mechanics?

Solve

Step 1: Write an expression for the regular cost of a tune-up and oil change.

$\$65 \times 1 \times 2 = \$130$ — \$65 per hour for 1 mechanic for two hours

Step 2: Subtract \$40 from the regular cost to find the cost on Thursdays.

Cost on Thursdays $= \$130 - \$40$
$= \$90$

Step 3: Use x for the rate that Shane charges customers per mechanic on Thursdays. Rewrite the expression in Step 1 with Thursday's numbers. Solve for x.

$x \times 1 \times 2 = \$90$ — One mechanic at \$$x$ per hour for two hours costs \$90.

$2x = \$90$ — Simplify. Solve for x.

$x = \$90 \div 2$ — Divide both sides by 2.

$x = \$45$

Step 4: Compare the per hour rate on Thursdays to the rate on other days. Subtract to find the difference.

$\$65 - \$45 = \$20$

Answer the Question

Step 5: The change in the markup per hour on Thursdays is \$20.

Now try these problems.

1. After school, students fix cars with the shop teacher's help. The students charge \$9 per hour. The average job takes 2 students working together 4 hours. Sometimes they work on Saturdays.

Then the cost is $30 more for the average job. What is the change in their hourly rate on Saturdays?

Answer: The auto repair shop students change their hourly rate

by $________ on Saturdays.

2. David has a computer repair business. He charges $30 per hour before 5 p.m. After 5 p.m., he charges $45 per hour. This week he works 28 hours before 5 p.m. He earns a total for the week of $1,110. How many hours does he work after 5 p.m.?

   A 6 **B** 15 **C** 25 **D** 39

3. During the summer, Magda runs a landscaping service. Because it is hot, she prefers to work only five hours a day, four days a week. Her schedule for the next week shows five customers need various jobs. How would you schedule these jobs during the week? Create a plan for each day.

Customers	Mow lawns	Trim bushes	Plant flower/veg beds
203 Twelfth Street	2 hours		
311 Station Street		1 hour	
19 C Street	4 hours		6 hours
621 Channel View		4 hours	
84 The Courtyard		2 hours	

4. Jaclann is a lifeguard. Every year, she renews her CPR certificate. Every 2 years, she renews her lifeguard certificate. Every 3 years, she renews her first aid certificate. This year she renewed all three. On the number line, mark C for CPR renewal, mark L for lifeguard renewal, mark F for first aid renewal. In how many years will she again renew all three in the same year?

This year

Answer: Jaclann renews the three certificates again in ______ years.

☆ *Challenge Problem*

You may want to talk this one over with a partner.

You run a travel agency. You get 10% commission on each airline ticket and 15% commission on each hotel reservation. One week you sell a total of $20,000 in airline tickets, and $2,500 in hotel reservations. How much commission do you earn? Explain how you got your answer.

Lesson 3 — Buying & Renting

Example Sigourney wants to buy a refrigerator for the office. The refrigerator costs \$499. She can pay for it all at once, or she can make monthly payments. Sigourney decides to make monthly payments. She puts \$100 down, and pays \$40 a month for 12 months. Which way of paying costs less? By how much?

Solve

Step 1: Calculate the amount of money Sigourney pays in 12 monthly payments.
$\$40 \times 12 = \$480$

Step 2: Add the down payment.
$\$480 + \$100 = \$580$

Step 3: Compare this amount of money with the cost of the refrigerator.
$\$580 - \$499 = \$81$

Answer the Question

Step 4: Paying all at once costs less by \$81. Credit buying costs her \$81.

✏ Now try these problems.

1. Xavier wants to buy a set of tools. The tools cost \$350. He makes monthly payments on a credit plan. He puts 25% down. Then he pays \$20 a month for 18 months. How much does he pay for the tools? Which way of paying costs more?

Answer: Xavier pays \$________ for the tools. ______________________ costs more.

2. Eliza rents a drywall screw gun. She pays \$11.55 per day. A drywall screw gun typically costs about \$150. How many days of rent would make it worth her while to buy one? Circle the correct answer.

A 12 **B** 13 **C** 15 **D** 20

3. Myron buys a lawn mower for $8,500. On average, he charges $25 to mow a yard.

   a. How many yards must he mow to recoup the cost of the mower?

   Answer: ________

   b. He mows four yards a day. What fraction of the cost of the mower does he recoup in one day? Shade the number of squares that reflect this fraction.

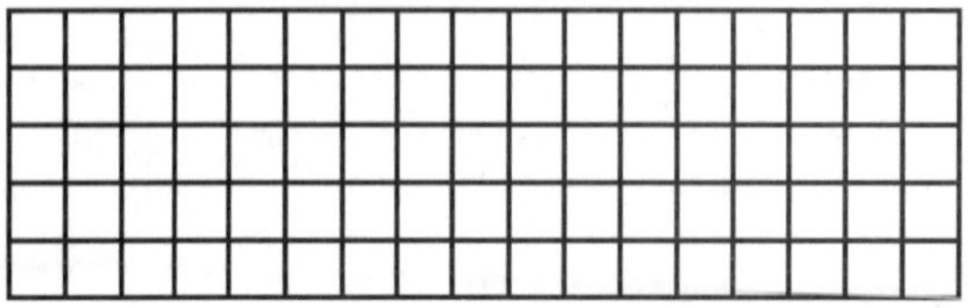

4. The Rent-to-Own store rents many items. Jon can rent a stereo system for $17 a week over 60 weeks. Or he can buy it for $527.00. Instead of paying rent, he saves $17 a week starting on January 1st. How much money does he save over the rental plan? Show the total of his savings by combining these bills. Under each denomination, write how many of that bill you will use.

__________ __________ __________ __________ __________ __________

☆ *Challenge Problem*

You may want to talk this one over with a partner.

You rent a machine to remove the vinyl flooring in your kitchen. The charge is $8 for the first hour and $4 for each hour (or part of an hour) afterwards.

a. You check out the machine at 8 a.m. You check it back in at 4:45 p.m. What do you pay?

____

b. It would have taken you about 60 hours to do the same job without the machine. Your time is worth $12 an hour. What is your true savings? Explain your answer.

____

____

____

____

Lesson 4

Selling & Sales

Example Madison is a sales person for a computer company. She earns a base salary of \$23,000 per year. She also earns a commission of 3% on all sales that she makes for the company. This year, she earns a total salary of \$35,000. What is the dollar value of the sales she makes?

Solve

Step 1: Subtract to find how much money Madison earns in commissions.

Total salary – base salary = Commissions

\$35,000 – \$23,000 = \$12,000

Step 2: Underline the sentence that tells what commission she earns on her sales.

<u>She … earns a commission of 3% on all sales that she makes for the company.</u>

Step 3: Write the same sentence using the language of math. Use x for the dollar value of all sales she makes for the company. Solve for x.

3% of x = \$12,000

x = \$12,000 ÷ 3% so x = \$12,000 × $(\frac{100}{3})$

x = \$400,000

Answer the Question

Step 4: The dollar value of the sales Madison makes is \$400,000.

Now try these problems.

1. Sergio is a sales person for *AllSports*. He earns a base salary of \$19,500 per year. He also earns a commission of 5% on all sales that he makes. This year, he earns a total salary of \$32,000. What is the dollar value of the sales he makes? Fill in the blanks to show the answer.

   Answer: The dollar value of the sales that Sergio makes is

   \$__________ ÷ 5% = \$__________.

2. A sports equipment manufacturer sells skateboards wholesale to *AllSports. AllSports* pays $55 for each skateboard when the order is for 100 or more boards. *AllSports* sells the skateboards at a retail price of $165 each. Which line entry in this list refers to skateboards? Check that entry.

Item code	Percent markup
16532	66.67%
87176	200%
91243	240%

3. On Saturday, 50 people visit the art gallery. Of these, 30% buy a piece of art. The gallery pays artists 60% of the price of their art when the art is sold. After paying the artists, the gallery makes $450 on Saturday. What is the mean price of a piece of art sold this Saturday?

A $18 **B** $30 **C** $75 **D** $180

4. Dario sells floor coverings. His customer compares carpeting to tiles for the floor in her family room. Carpeting costs $65 per square yard. Tiles cost $9 per square foot. Which floor covering will cost more? Explain why. Use a diagram to support your explanation.

Answer: ______________________________

☆ *Challenge Problem*

You may want to talk this one over with a partner.

A ranch owner has 6,315 steers. He gives one third of them to his son. He gives 20% of the rest to his cousin. He sells $\frac{3}{4}$ of the rest to another rancher. How many does he have left? Explain your answer.

____

____

Lesson 5

Cross Numbers

A Game (for One or Two Players)

The goal of this game is to complete the *Cross Numbers* puzzle. First, you read each clue. Then, you figure out the answer and write it in the related spaces.

Materials

Game board (on the next page), pencils

Directions

1. Players can play alone or against each other. When a player plays alone, he or she can play against the clock.

   When two players play, one can answer the odd-numbered clues, and the other can answer the even-numbered clues.

   In another way to play, one player can answer the *Across* clues, and the other can answer the *Down* clues.

2. When two players play, they alternate turns.

   In turn, each player reads a clue and writes the answer in the related spaces. You may have to put a zero in front of the answer to make it fit the space. When a player disagrees with a number that is already written in a particular space, both players check their answers. The player with the incorrect answer erases his or her answer.

3. The winner is the first player to complete the *Cross Numbers* puzzle with correct answers.

Before you play the game, try these warm-up problems.

1. You play *Cross Numbers* with a friend. You complete 6 of the 7 *Down* clues. Your friend completes 6 of the 8 *Across* clues. Who has completed the larger percent of their clues? By how much?

   A You, by 11% **C** Your friend, by 11%

   B You, by 8.7% **D** Your friend, by 8.7%

2. You and a friend roll a number cube to see who will go first in *Cross Numbers*. What is the probability that both of you will roll a 5?

   Answer: The probability is ______.

Game Board for *Cross Numbers*

1	2		4	5	6
7				8	
9					
		10			
11				12	
			13		

Across	Down
1. One quarter of your $4,008 monthly income pays for office furniture. What does the furniture cost?	1. 60% of the cost of a rocking chair is labor. One chair costs $2,500. What is the cost of labor?
5. You pay $33 to rent a dolly for three days. What is the mean rental cost per day?	2. *SK!* buys skateboards for $50, and sells them for $144 each. What does *SK!* make on 3 boards?
7. You and a friend share the rent on storage space equally. What percent do you each pay?	4. You get 5% commission on all sales. Your sales are $4,240. What is your commission?
8. You take 2.5, 4.5, and 2 hours on three occasions to fix mowers. What is the mean time per mower?	5. You cut 4-inch squares out of 1 yard of 48-inch wide fabric. How many squares can you cut?
9. You put $300 down on a $2,500 computer. You pay $146 a month for 18 months. What does the computer cost you?	6. You plant a strip of flowers all around the 3-meter-by-4-meter toolshed, except in front of the 1-meter-wide door. How long is the whole strip?
11. Artists pay the art gallery 40% of each sale. Your art sells for $200. How much do you get?	10. *TempS* pays you $10 per hour. They charge you out at $15 per hour. What is the percent markup?
12. You charge $30 an hour on weekdays. On weekends you charge $45. How many weekend hours earn you as much as 21 weekday hours?	12. You build birdhouses with 8-inch by 6-inch bases. How many bases can you cut from a 2-foot square piece of wood?
13. You save $20 a week for a TV that costs $240. In how many weeks can you buy it outright?	

Unit 2 Review

Review What You Learned

In this unit you have used mathematics to solve many problems. You have used mental math and estimation, practiced basic operations, solved equations, used statistics and probability, and used measurements on figures of different shapes with both standard and metric measurements.

These two pages give you a chance to review the mathematics you used and check your skills.

✔ Check Your Skills

1. Maria chooses tiles for the floor of her 4-meter by 3-meter bathroom. The tiles are $\frac{1}{2}$-meter squares. How many tiles does she need? Draw a diagram. Label the dimensions and number of tiles along each side.

Answer: Maria needs ______ tiles.

If you need to review, return to lesson 1 (page 15).

2. After school, students run a carpentry shop. The students charge $7 per hour. The average job takes 2 students each working 6 hours. Sometimes they work on Saturdays. Then they charge $30 more for the average job. How much more per hour is the rate on Saturdays? Complete the steps. Find the answer.

Cost of average job during the week = $7 × _____ × _____ = $_____

Cost of average job of Saturdays = $_____ + $_____ = $_____

Rate that students charge on Saturdays = $_____ ÷ ________ = $_____

Increase in hourly rate on Saturdays = $_____ – $_____ = $_____

If you need to review, return to lesson 2 (page 17).

3. Georgeanna wants to buy a copier for the office. The copier costs $599. She can pay for it all at once, or she can make monthly payments. She makes monthly payments. She puts $100 down, and pays $25 a month for 24 months. Which way of paying costs more? By how much?

Answer: ______________________________ costs more.

The difference is $______.

If you need to review, return to lesson 3 (page 19).

4. Pyn rents a heavy-duty sewing machine to sew sails. He pays $12 the first day and $8 a day afterwards. A similar heavy-duty sewing machine costs $348. How many days of rent would make it worth his while to buy one?

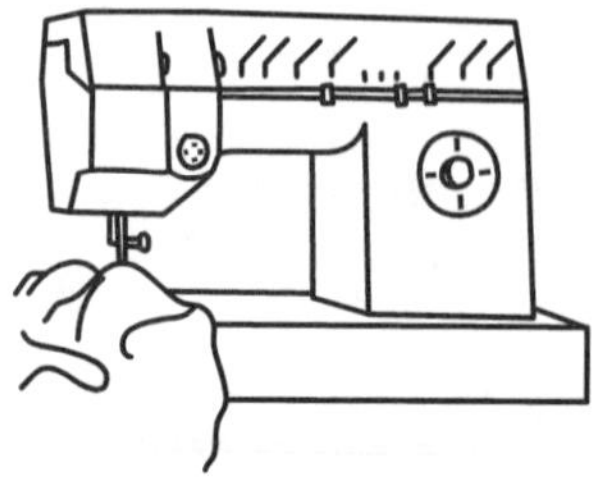

A 17.4 **B** 29 **C** 42 **D** 43

If you need to review, return to lesson 3 (page 19).

5. Brett sells farm equipment. He earns a base salary of $21,000 per year. He also earns a commission of 6% on all sales that he makes. This year he sells $260,000 of equipment. What is his total salary, including commission? Write an expression and find the answer.

Answer: ______________________________ = $__________

If you need to review, return to lesson 4 (page 21).

Write Your Own Problem

Choose a problem you liked from this unit. Write a similar problem using a situation and related facts from your own life. With a partner, share and solve these problems together. Discuss the mathematics and compare the steps you used. If you need to, rewrite or correct the problems. Write your edited problem and the answer here.

____

____

____

Unit 3

Community Service

Preview

How You Will Use This Unit

Communities provide many opportunities for service. You can help paint or repair buildings around town. You can deliver meals or visit people in the hospital. You can take part in town clean-up events. You can lend a hand with a latchkey program or at an animal shelter. As you participate, you will often use math. The math skills you use include mental math and estimation, basic operations and equations, ratios and proportions, statistics and probability, and various measurements (standard and metric).

What You Will Do in This Unit

In this unit, math steps demonstrate how to solve problems. These steps can help you answer questions such as these:

You prepare to seed the retirement home lawn. The lawn is 50 feet by 30 feet. One pound of seed covers 300 square feet (if all seeds sprout). How many one-pound bags of seed do you buy?

Once a week, you visit a local children's hospital. Four out of five times, the children want to play games. What is the theoretical probability that today the children ask for a book?

In the past, one out of six people volunteer to help with the beach clean-up projects. You want at least 50 people to help with the next project. How many people do you need to ask?

You report that 12 turtles make up 24% of the total animal population at the shelter. How many animals are there at the shelter?

What You Can Learn from This Unit

When you complete this unit, you will have used mathematics to work problems related to community service. These problems are similar to those that may actually occur in your daily life.

Lesson 1 → *Building Community*

Example The Science Club prepares to seed the retirement home lawn. The lawn is 50 by 30 feet. If all the seeds sprout, you could cover 300 square feet with one pound of *Greenway* grass seed However, the nursery says that about 90% of the seeds will sprout. So you will need more than one pound to cover 300 square feet. How many one-pound bags of seed does the club buy?

Solve

Step 1: Calculate the area of the retirement home lawn.

50 feet × 30 feet = 1,500 square feet

Step 2: Underline the sentence that tells the area that one pound of seed covers.

Underline the sentence that tells the percent of seeds that will sprout.

One pound of *Greenway* grass seed covers 300 square feet (if all seeds sprout).

The nursery guarantees that 90% of the seeds will sprout.

Step 3: Combine the information from the two underlined sentences. Write a sentence to tell what area the seeds that sprout will cover.

One-pound of this seed will cover 90% of 300 square feet, or 270 square feet.

Step 4: Write a math expression for the number of bags of seed needed to cover the Retirement Home lawn.

Number of one-pound bags of seed = [1,500 ÷ 270] square feet
= 5.56

Answer the Question

Step 5: The Science Club buys 6 one-pound bags of seed.

✏ Now try these problems.

1. On Make-A-Difference Day, the shop class re-gravels the theater parking lot. The parking lot is 80 feet by 30 feet. In theory, one wheelbarrow load of gravel covers 25 square feet. The class finds

that their gravel actually covers only 80% of that area How many wheelbarrow loads of gravel does the shop class need?

Answer: The shop class needs ________ wheelbarrows of gravel.

2. Tom's art students design a mural for City Hall. The mural is 12 feet square. Twenty four students sign up to paint the mural. Tom divides the mural into 24 equal-size panels. On the day of painting, 2 students are sick and cannot paint. What size area does Tom need to reassign?

   A 1 square foot **C** 6 square feet

   B 2 square feet **D** 12 square feet

3. By mistake, Ellis spills some paint in the driveway. Jake drives across the paint on his bike. He gets paint on 7 inches of his front wheel. The wheel has a diameter of 28 inches. What paint pattern does he leave in the driveway over the next few yards? Draw a diagram. Write the measurements on the diagram. ($C = \pi d$)

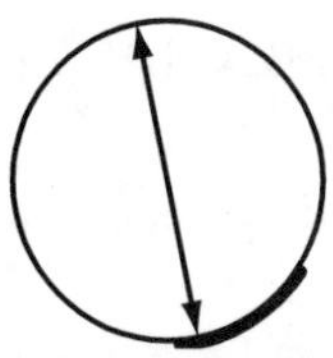

4. Yancey builds a circular booth to display community messages. It stands 2.5 meters tall. The circumference measures 2.25 meters. Draw a line to match each description to the correct dimension. ($A = \pi r^2$; $A = lw$; $C = 2\pi r$)

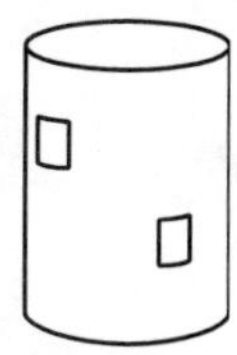

Area of kiosk's base	Area of display space	Number of 22-cm by 17-cm flyers that the booth can accommodate
$(225 \times 250) \div 374$	(2.25×2.5) square meters	$(2.25)^2 \div 4\pi$ square meters

☆ *Challenge Problem*

You may want to talk this one over with a partner.

You conduct a survey to find out which of 16 issues are important to young people. The city conducts a survey to find out which of 27 issues are important to the rest of the community. The two surveys have twelve issues in common. How many issues are unique to each survey? What kind of graph or diagram would you use to help to show your answer? What kind of graph or diagram could help you communicate the results of the survey?

____

____

Lesson 2

Brightening Someone's Day

Example Once a week, Kim and his two younger sisters visit a local children's hospital. They take books to read. But four out of five times, the children want to hear about school or play games. Today, Kim and his sisters go straight from school with no books. What is the theoretical probability that the children ask for a book?

Solve

Step 1: Underline the sentence that tells how many times Kim and his sisters don't need books.

But four out of five times, the children want to hear about school or play games.

Step 2: Now, write the number times out of five that they need a book.

1 out of 5 times

Step 3: Write the theoretical probability that the children ask for a book.

$\frac{1}{5}$ $\qquad$ $\text{Probability} = \frac{\text{Favorable outcomes}}{\text{Total possible outcomes}}$

Answer the Question

Step 4: The theoretical probability that the children ask for a book is $\frac{1}{5}$.

Now try these problems.

1. On Wednesdays, Margaret visits an elderly lady on her street. One out of six times, the elderly lady does not need Margaret to vacuum and dust. What is the theoretical probability that Margaret will vacuum and dust for the elderly lady?

   Answer: The theoretical probability is ______.

2. The youth club spends one whole day fixing things for the elderly and disabled. They ask companies or individuals to sponsor the

event on a \$1/hour/student basis. The students record their hours.

Number of students	Number of hours per student
7	9
5	6
4	5
9	3

a. What is the total amount of money that they can collect from sponsors? What is the mean number of hours worked per student?

A \$25; 5 hours

B \$23; 6.1 hours

C \$140; 5.6 hours

D \$575; 5.75 hours

3. On Thursdays, Pat delivers meals to people who cannot prepare food for themselves. This month, the median of the number of hours that she donates is 3 hours a day. The range is 2.5 hours. What is the least and the most number of hours a day that she donates? Mark each number on the number line.

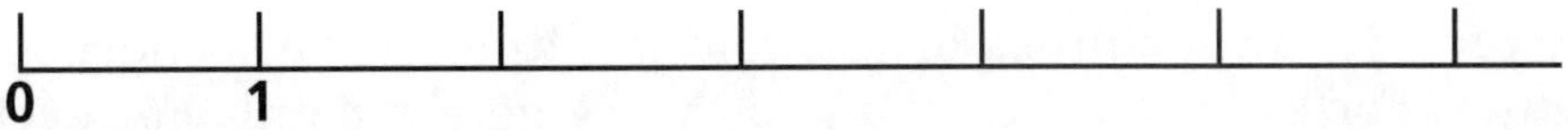

4. Paul volunteers to work an extra day at the library so Celeste can help her mother. He gets there a half-hour early and stays a quarter-hour late. Celeste normally works 6 hours. How many hours does Paul work?

Answer: ___________

☆ *Challenge Problem*

You may want to talk this one over with a partner.

You run a food drive to collect food. Then you box it up and take it to families in need. To cut costs, you buy large 90-inch by 108-inch sheets of cardboard. You cut out open boxes, each 18-inch square, from the cardboard. Circle the box design that wastes the least cardboard. Draw a diagram to show a layout on the cardboard to make 6 boxes.

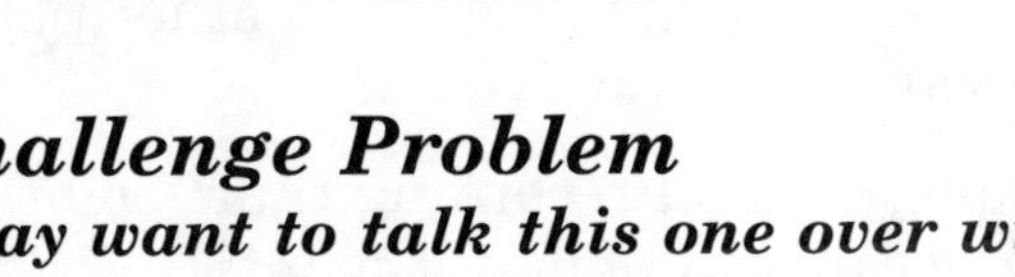

A **B** **C** **D**

Lesson 3

Community Space

A Pegboard Game (for One or More Players)

The goal of this game is to construct polygons of different sizes. First, you construct a polygon. Then, you challenge the next player to construct a polygon using one fact about your polygon.

Materials

Game board (on the next page), rubber bands of different colors. This game can be played with a geoboard of any size, such as 5×5 or 11×11.

Directions

1. Pick a rubber band color for each player. Count the boundary pegs. A boundary peg touches the rubber band.
2. Player 1 constructs a polygon on the pegboard using a rubber band. (For example, the polygon shown on the next page.) The number of pegs on the boundary, *b*, divided by 2, plus the number of pegs inside, *i*, minus 1, is equal to the area, *A*. In the triangle shown on the next page, $b = 10$ and $i = 6$. So $A = \frac{b}{2} + i - 1$ and $A = 10$. Check the area of the triangle by using the formula, $A = \frac{1}{2} \times \text{base} \times \text{height}$.)
3. Next, player 1 tells the values of *b, i,* and *A*. Other players check these values. For every correct value, player 1 gets a point. For every incorrect value, player 1 loses a point. Player 1 then chooses one of the variables *b,* or *i,* or *A,* for Player 2.
4. Player 2 constructs a different polygon that has the same *b, i,* or *A,* whichever is chosen by player 1. For example, suppose player 1 makes the triangle shown, and then chooses the *A* (which is 10). Then player 2 constructs a polygon that has an area of 10 (for example, a rectangle that is 2×5). Player 2 tells the new values of *b, i,* and *A,* with other players checking. Player 2 then chooses a variable, *b,* or *i,* or *A*.
5. Players take turns constructing polygons and recording their points. The winner is the player with the greatest number of points.

Before you play the game, try these warm-up problems.

1. You construct a polygon with 12 boundary pegs and 4 inner pegs. Circle the correct area.

   9 square units   12 square units   14 square units   16 square units

2. Carlos says his polygon has 16 boundary pegs, 3 inner pegs, and an area of 12 square units. Is he correct? If not, what could he have done wrong?

   Answer: ______________________________

Game Board for *Community Space*

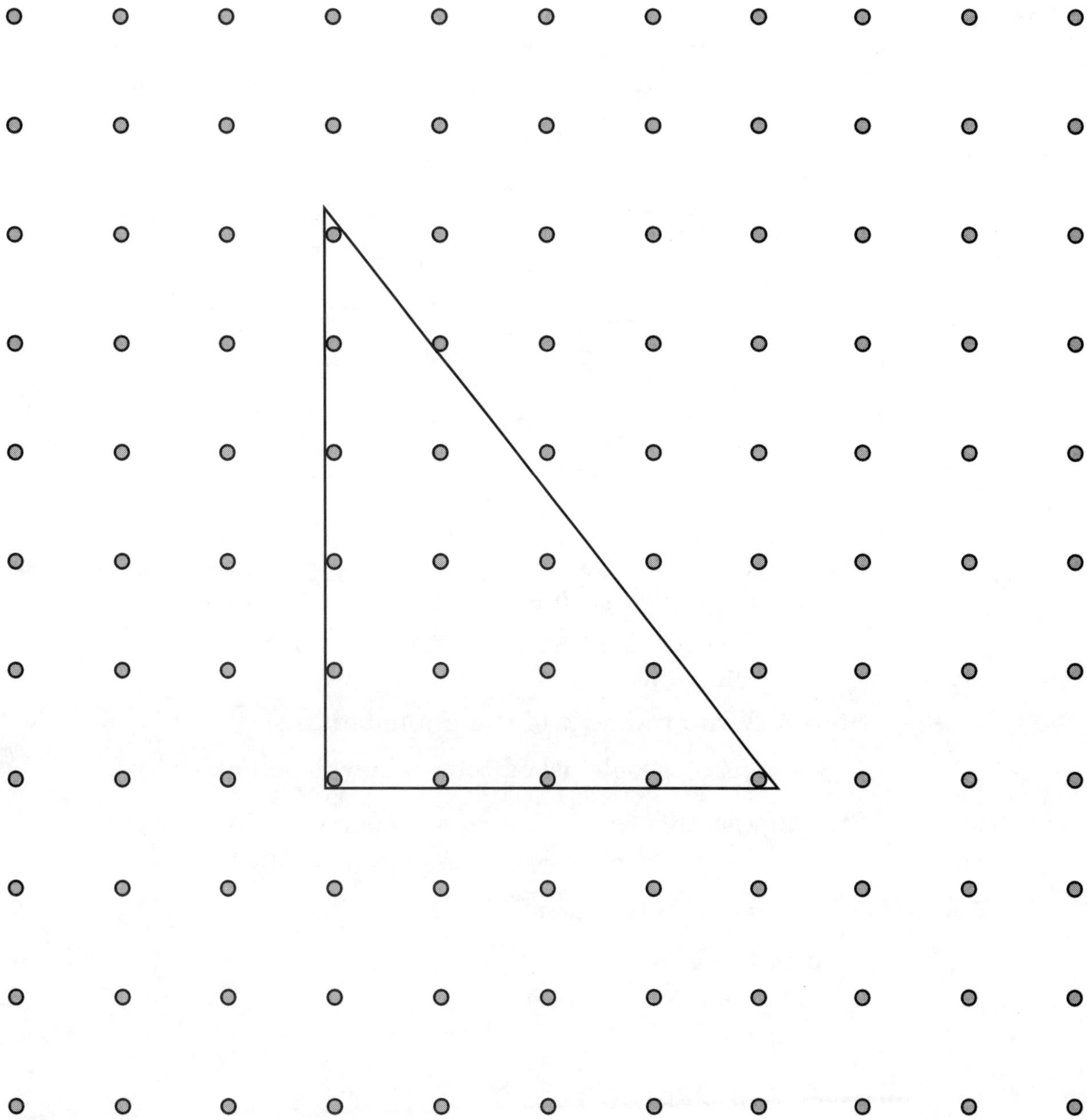

Lesson 4 — Cleaning Up

Example About one out of every six people asked volunteer to help with beach clean-up projects. *OurBeach* wants at least 50 people to help with the next project. How many people do they need to ask?

Solve

Step 1: Underline the sentence that compares the number of people asked to the number of volunteers.

<u>About one out of every six people asked do volunteer ...</u>

Step 2: Write the ratio of these numbers.

$\frac{6}{1}$ people asked to people who volunteer

Step 3: Use this ratio of write a proportion. Use x for the number of people *OurBeach* needs to ask.

$\frac{6}{1} = \frac{x}{50}$

Step 4: Solve for x.

$x = 6(50)$
$x = 300$

Answer the Question

Step 5: *OurBeach* needs to ask 300 people to get 50 volunteers.

Now try these problems.

1. Every year, the ecology club asks for students to help organize Earth Day events. Usually one student out of every four students volunteers to help. This year, the club needs at least 12 students to help. How many do should they ask?

   A 12 **B** 16 **C** 36 **D** 48

2. This year, the *TrashOff!* plans to clean up a future nature preserve. Lance divides the preserve into lots. The area of a lot is based on the size of the volunteer group. What size are the lots for groups with 11–15 people, and 16–20 people?

Group size	Lot size	Group size	Lot size
1–5 people	875 square meters	11–15 people	
6–10 people	1,750 square meters	16–20 people	

3. Ellese reviews the community's participation at the Great Community Mulch event. She ranks the streets according to their participation. How does she decide the ranks?

Street name	Number of residences	Participating residences	Rank
C Street	432	108	2
11 Avenue	720	144	3
Pecan Drive	54	18	1
Mountain Vista	102	17	4

Answer: ______________________________

4. Guiseppe buys coffee beans to make coffee for volunteers after the project. One pound of coffee beans makes 5 cups of ground coffee. One cup of ground coffee makes 32 ounces of brewed coffee. How many 8-ounce cups of brewed coffee can he make from one pound of coffee? Use the symbols, +, –, ×, ÷ to fill in the blanks and complete the equation.

Answer: Number of 8-ounce cups of brewed coffee = 5 __ 32 __ 8 = ____

☆ *Challenge Problem*

You may want to talk this one over with a partner.

You are standing on a trail in a local park. You see a friend walking on another trail that intersects the trail you are on. You are both the same distance from the intersection. You walk towards the intersection. Your friend walks at the same rate as you, but away from the intersection. When are the two of you closest? Explain your answer.

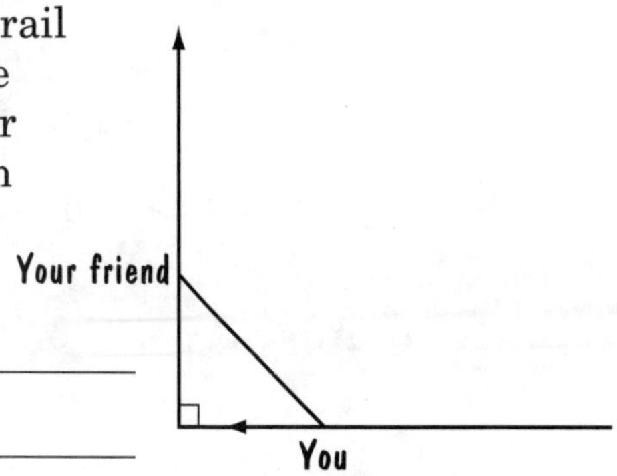

____

____

Lesson 5 — *Lending a Hand*

Example The volunteers at the marine-animal shelter make status reports. Lera makes a pie chart to show the relative percentage of each type of animal in the shelter. There are 12 turtles. How many of each of the other types of animal is there?

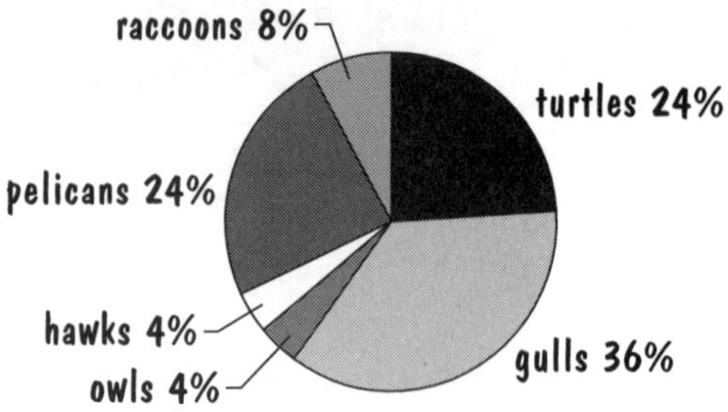

Solve

Step 1: Relate the number of turtles to the percentage of turtles.

12 turtles make up 24% of the total animal population.

Step 2: Write the same sentence using the language of math. Use t for the total number of animals. Solve for t.

$12 = \frac{24}{100} \times t$

$t = 12 \div (24 \times 100)$

$t = 50$

Step 3: Now, find the number of each of the other types of animals.

Number of gulls: $\frac{36}{100} \times 50 = 18$.

Number of owls: $\frac{4}{100} \times 50 = 2$.

Number of hawks: $\frac{4}{100} \times 50 = 2$.

Number of pelicans: $\frac{24}{100} \times 50 = 12$.

Number of raccoons: $\frac{8}{100} \times 50 = 4$.

Answer the Question

Step 4: There are 18 gulls, 2 owls, 2 hawks, 12 pelicans, and 4 raccoons.

Now try these problems.

1. Use Lera's pie chart. Lera checks her animal counts. There are only 11 turtles. There are more gulls. The numbers and percentages for all the other animals stay the same.

a. What can you say about the total number of animals at the shelter? Why?

Answer: a. There are ______ animals at the shelter because

____

b. How many gulls are there?

Answer: b. There are ______ gulls there.

2. On September first, the pet store has 11 iguanas which are 22% of the total animal population. Then 2 iguanas are released. (There is no change in the numbers of the other animals.) What percent (p) of the new total animal population (n) are the remaining iguanas? Fill in the blanks to find the answer. Explain your reasoning.

__________ represents the total number of animals before the release.

11 = ______ /100 of ______

n = ______

9 = ______/100 $\times$ n

p = ______

The remaining iguanas are ______ of the new total animal population.

3. Craig volunteers at the information desk in the local hospital. He works 3 hours an evening, 2 evenings a week. How many hours does he contribute in a year (of 52 weeks)?

A 24 **B** 104 **C** 156 **D** 312

4. Three volunteers use their cell phones to make calls to increase club membership. Cell phone charges include a base rate plus a cost per minute. What is the base rate? What is the cost per minute?

Name	Number of Minutes	Total Charge
Dana	25	$ 7.25
Hans	1 hour 15 minutes	$10.75
Vernette	1 hour 35 minutes	$12.15

☆ *Challenge Problem*

You may want to talk this one over with a partner.

You volunteer as a latchkey aide. According to the director, the number of children equals nine more than two thirds the number of children. How many children are in the program? Write an equation and find the answer.

____

____

Unit 3 Review

Review What You Learned

In this unit you have used mathematics to solve many problems. You have used mental math and estimation, ratios and proportions, statistics and probability, various measurements (standard and metric), practiced basic operations, and solved equations.

These two pages give you a chance to review the mathematics you used and check your skills.

✔ Check Your Skills

1. The Garden Club prepares to seed the gardens at the park. The gardens cover 1,700 square feet. If all the seeds sprout, you could use one pound of wildflower seeds for 250 square feet. The nursery guarantees that 95% of the seeds will sprout. How many one-pound bags of seed do they buy to cover half the garden area? Circle that number of bags.

If you need to review, return to lesson 1 (page 28).

2. Once a week, Coreen shops for an elderly neighbor. Six out of seven times, he asks if she remembered to buy his favorite cookies. What is the theoretical probability that today he does *not* ask? Circle the best answer.

   $\frac{1}{7}$ $\qquad$ $\frac{1}{6}$ $\qquad$ $\frac{6}{7}$ $\qquad$ $\frac{7}{6}$

   If you need to review, return to lesson 2 (page 30).

3. Usually one out of every three people who are asked, volunteer for community projects. The town wants at least 30 people to help with the next town clean-up project. How many people do they need to ask?

   Answer: ______

   If you need to review, return to lesson 4 (page 34).

4. Grandis buys coffee beans to make coffee for volunteers after the project. One pound of coffee beans makes 5 cups of ground coffee. One cup of ground coffee makes 32 ounces of brewed coffee. How many 6-ounce cups of brewed coffee can she make from two pounds of coffee?

A $26\frac{2}{3}$ **B** $53\frac{1}{3}$ **C** 160 **D** 480

If you need to review, return to lesson 4 (page 34).

5. Paloma's report shows that 8 puppies make up 10% of the total animal population at the shelter. Her next report shows that puppies make up only 5% of the total. She also says that the total number of animals at the shelter has increased by 20. How many puppies are there now? What is the total number of animals at the shelter?

Answer: There are now ______ puppies at the shelter.

The total number of animals is ______.

If you need to review, return to lesson 5 (page 36).

Write Your Own Problem ✍

Choose a problem you liked from this unit. Write a similar problem using a situation and related facts from your own life. With a partner, share and solve these problems together. Discuss the mathematics and compare the steps you used. If you need to, rewrite or correct the problems. Write your edited problem and the answer here.

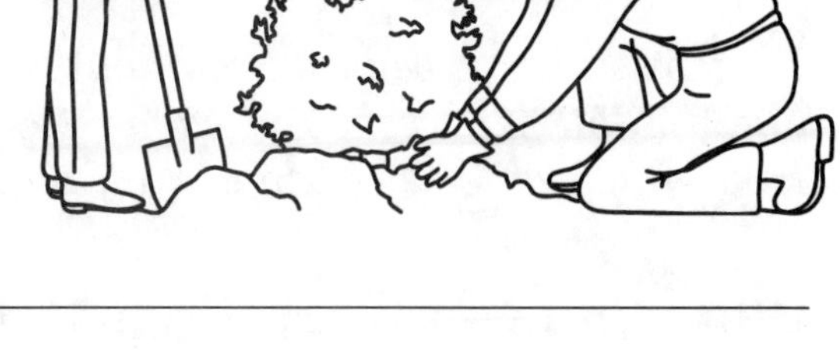

____

____

____

____

____

Unit 4

Profit & Loss

Preview

How You Will Use This Unit

A key question in running a business is whether you are making a profit or a loss. You compare income with expenses, and assets to liabilities. If you start up a new business venture, you consider all the start-up and marketing costs. From time to time, you may also make donations. As you review your financial situation, you will often use math. The math skills you use include mental math and estimation, basic operations and equations, and statistics and probability.

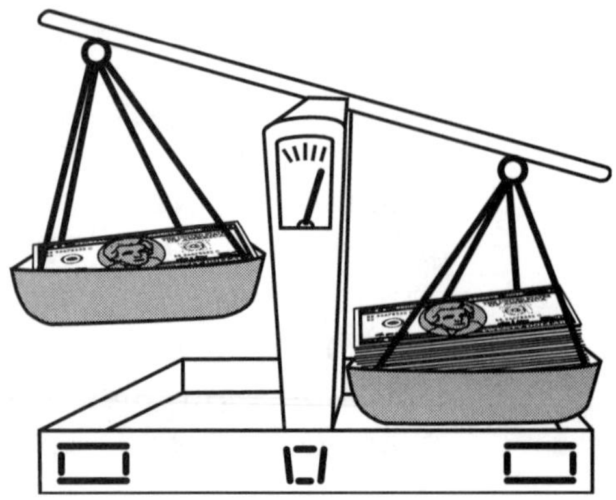

What You Will Do in This Unit

In this unit, math steps demonstrate how to solve problems. These steps can help you answer questions such as these:

In the first three months of the year, you spend $2,500, $3,450, and $1,850. Your income is $2,500 a month. What can you spend per month for the rest of the year?

You review your net worth. Assets include property, such as a truck or a piece of jewelry. Liabilities include financial obligations, such as a loan or bills. What is your net worth?

You earn $21,400 a year. Your expenses total about $1,600 per month. In addition, you donate $50 to a charity. What effect does this donation have on the taxes you pay?

Your business expenses are about $650 per week. Your living expenses are about $350 per week. You earn $1,050 a week for your design services. How much is left over to pay off start-up costs?

What You Can Learn from This Unit

When you complete this unit, you will have used mathematics to work problems related to business and profit versus loss. These problems are similar to those that may actually occur in your daily life.

Lesson 1

Income & Expenses

Example In the first four months of the year, Cassandra spends \$4,500, \$5,850, \$2,125, and \$1,850. She does not want to exceed \$3,000 a month in expenses over the year. What is the average (mean) amount she can spend per month for the rest of the year?

Expenses for the Current year	
January	\$4,500
February	\$5,850
March	\$2,125
April	\$1,850

Solve

Step 1: Calculate the maximum that Cassandra can spend this year.

$\$3,000 \times 12 = \$36,000$ Multiply the mean amount per month by the number of months in a year.

Step 2: Find the total amount she has spent so far this year.

$\$4,500 + \$5,850 + \$2,125 + \$1,850 = \$14,325$

Step 3: Calculate the mean per month for the months remaining.

$\$36,000 - \$14,325 = \$21,675$ Subtract amount spent from amount available.

$\$21,675 \div 8 = \$2,709.38$ Divide by the number of months left in the year.

Answer the Question

Step 4: The mean amount that she can spend per month is \$2,709.38.

Now try these problems.

1. In the first five months of the year, Tate spends \$1,940, \$4,230, \$1,600, \$2,750, and \$1,850. He can afford no more than \$30,000 in expenses this year. What is the mean amount that he can spend per month for the rest of the year? Is this above or below his monthly goal? By how much?

   Answer: Tate can spend \$__________ per month. This is

   __________ (above/below) his goal by \$__________.

2. Zack goes to the movies at least once a week, for all 52 weeks in a year. He spends \$7 on a ticket, \$5 on popcorn and soda, and

$10.50 on pizza afterwards. How much does this add up to over the year? Circle the correct answer.

A $22.50 **B** $270 **C** $1,170 **D** $2,366

3. Marshall reviews his quarterly income and expenses for his bicycle repair business. He expects his expenses in the 4th quarter of the year to be the same as the 3rd quarter. How much income must he make in the 4th quarter to meet his total year's expenses? Fill in the line entries in the chart. Draw bars on the bar chart.

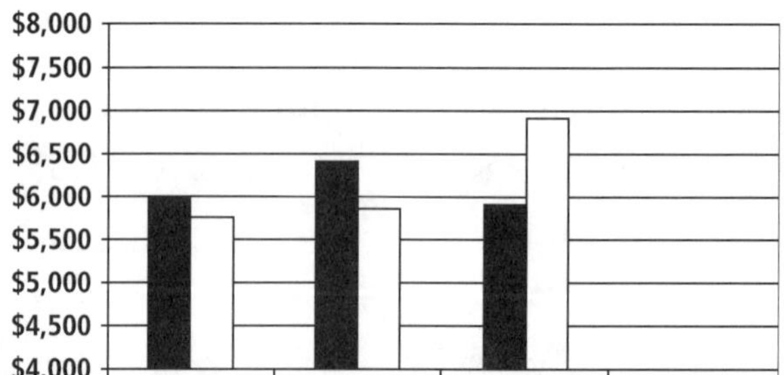

	1st	2nd	3rd	4th
■ Income	$6,000	$6,400	$5,900	
□ Expenses	$5,750	$5,850	$6,900	

4. Last year, to cover expenses, the theater rented out booths at *BeachFest* for $15 each. This year, they instead charge a fee of 5% of all goods sold. There are 50 booths. The mean amount of goods sold per booth is $400. Which method makes more income for the theater? What can change the picture?

Answer: ____

☆ *Challenge Problem*

You may want to talk this one over with a partner.

What do you think your living expenses are per month? You take an unplanned-for vacation that costs $2,736. How much surplus income do you need per month to meet this one-time expense in your annual budget? Explain your answer.

____

____

Lesson 2

Profit or Loss

A Card Game (for Two or More Players)

The goal of this game is to get the greatest profit and the least loss. First, you take a *profit-or-loss* card. Then you answer the question and update your *balance line*.

Materials

Profit-or-loss card deck (on the next page), blank *balance line* sheets (on next page).

Directions

1. Place a blank balance line sheet in front of each player. Place the shuffled *profit-or-loss* card deck of 40 cards face down in the center of the playing area.

2. Player 1 picks a *profit-or-loss* card from the card deck. The player calculates the value of the expression on the card. Then the player writes the value above or below the *balance line* on the player's card. (If the value is positive, the amount goes above the *balance line*. If it is negative, it goes below the *balance line*.) Other players check the value. The *profit-or-loss* card is then placed in the discard pile.

3. Players take turns taking *profit-or-loss* cards and updating their *balance lines*. When all the cards are gone, shuffle the discard pile and use it as the new card deck.

4. The game is over when a player picks one of the (four) cards. To find the winner, each player subtracts the loss amount from the profit amount on the player's card. The winner is the player with the greatest positive result.

Before you play the game, try these warm-up problems.

1. Sandi draws a *profit-or-loss* card that says 50% of 50. What does she write on her *balance line* sheet?

   Answer: She writes ______ (value) on her sheet __________ (above/below) the line

2. So far, Danhui has 620 above the line and 496 below the line on his *balance-line* sheet. Canta has 570 above the line and 445 below the line on his *balance-line* sheet. The next card is a . Who wins the game and by how much?

   Answer: ___________ wins the game by ___________________________.

Profit or Loss Game

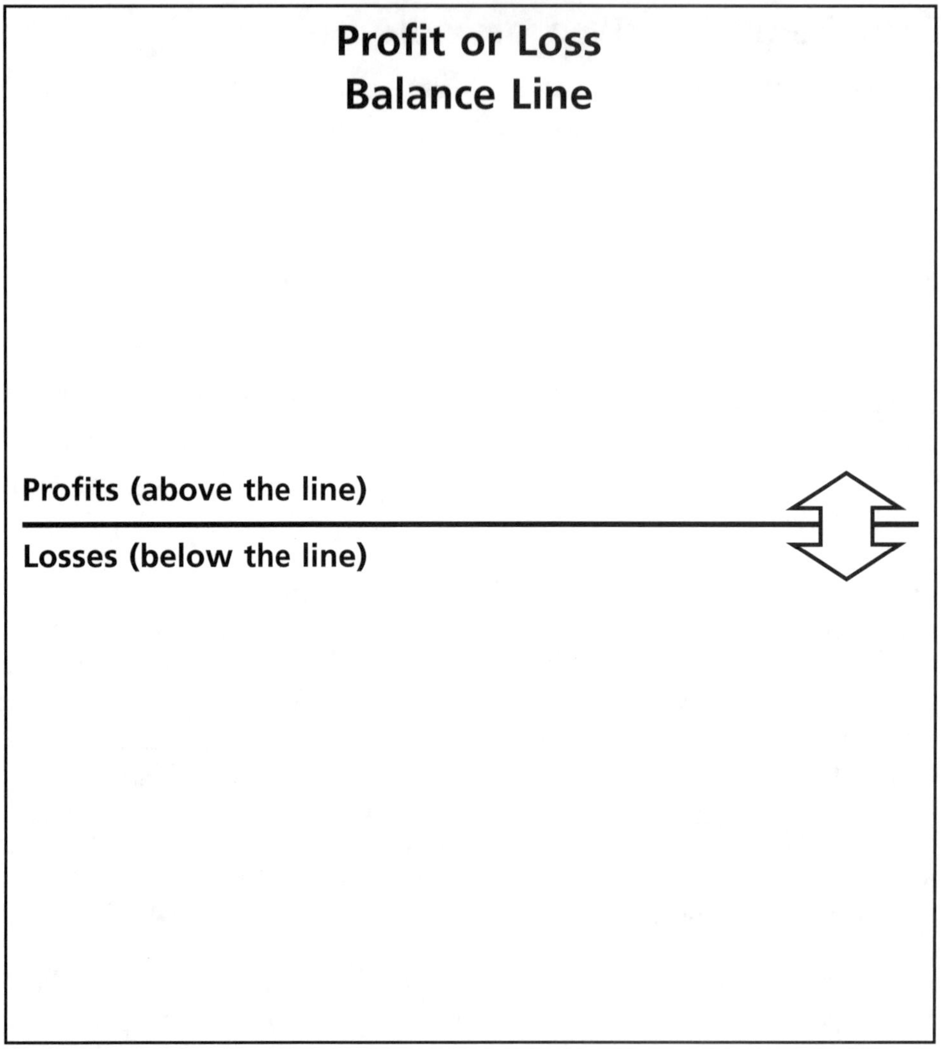

Profit-or-Loss Cards

Write each expression on a card, one to a card, to make 36 cards.
To complete the deck of 40, make 4 cards that have ⇕ on them.

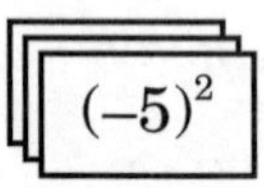

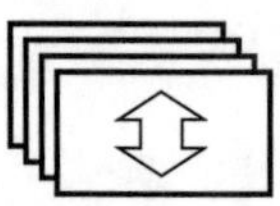

$80 + 27$	$16 + 24$	$21 + 15$	$14 + (-5)$
$35 - 64$	$42 - 17$	$-(29 - 18)$	$-(19 - 87)$
$\frac{85}{5}$	$\frac{102}{6}$	$\frac{\sqrt{81}}{2}$	$-\frac{\sqrt{49}}{3}$
13×6	7×9	$(-6) \times (-20)$	$(-5) \times 4$
50% of 50	45% of 100	−(8% of 150)	30% of (−90)
$\frac{1}{2}$ of 28	$\frac{1}{3}$ of 900	$\frac{1}{4}$ of 148	$-\frac{1}{5}$ of 250
$\sqrt{6400}$	$\sqrt{625}$	$-\sqrt{3600}$	$-\sqrt{16}$
$(-5)^2$	$(-9)^2$	$(3)^2$	$(2)^2$
$(-2)^3$	$(-3)^3$	$(5)^3$	$(2)^3$

Lesson 3

Assets & Liabilities

Example Billy reviews his net worth. Net worth is the difference between assets and liabilities. Billy's assets include property, such as his truck and money available in a bank account. His liabilities include repaying the loan from his uncle, and paying current bills. What is his net worth?

Balance Sheet Summary		
Assets		
Truck and trailer	$18,000	
Kayaks	$14,000	
Bank balance	$ 3,500	
Total	**$35,500**	
Liabilities		
Mortgage on garage	$21,600	
Uncle's loan	$ 5,000	
Current bills	$ 2,200	
Total	**$28,800**	
NET WORTH		**$6,700**

Solve

Step 1: Add the value of Billy's assets. Write the answer on the asset total line.

$18,000 + $14,000 + $3,500 = $35,500

Add the value of Billy's liabilities. Write the answer on the liabilities total line.

$21,600 + $5,000 + $2,200 = $28,800

Step 2: Subtract liabilities from assets to find Billy's net worth. Write the answer on the net worth line.

$35,500 – $28,800 = $6,700

Answer the Question

Step 3: Billy's net worth is $6,700.

✏ Now try these problems.

1. Taliatha's assets include quilting equipment and supplies. Her liabilities include credit card bills and other bills. Her assets total $4,300. Her liabilities total $10,630. What is her net worth? Complete the equation with one of these symbols: +, –, ×, or ÷. Find the answer.

   Answer: Taliatha's net worth is $4,300 ___ $10,630 = $_________.

2. At the first of the year, Stanton's assets total $2,450. His liabilities total $3,400. This semester, his father gives him a $8,400 truck. He also takes out a $4,000 loan to go to school.

Draw three arrows to show the numbers giving (A) his net worth at the first of the year, (B) his new net worth, and (C) the increase in his assets .

$3,450	**$7,400**	**$10,850**
–$6,450	**$8,400**	**$1,600**
$2,450	**$3,400**	**–$950**

3. Debbie wants to calculate the average yearly depreciation on her car. She subtracts the price for which she sells the car from the price she paid for it. Then she divides by the number of years she has owned it. Debbie paid $12,500 for the car. Four years later, she sells it for $7,200. What is the average yearly depreciation?

A $1,325 **C** $3,125

B $1,800 **D** $19,700

4. Dean uses this chart to compare three different investments. Use the formula, $I = p \times r \times t$ to find the missing values.

Interest (I)	Principal (p)	Annual Rate (r)	Time (t, in years)
$__________	$400	5%	2 years
$120	$__________	6%	1 year
$150	$2,000	______	3 years

☆ *Challenge Problem*

You may want to talk this one over with a partner.

Each of the three kinds of cube is assigned a different number value. Find a set of numbers that makes the sum of the cubes on each side of this equation equal.

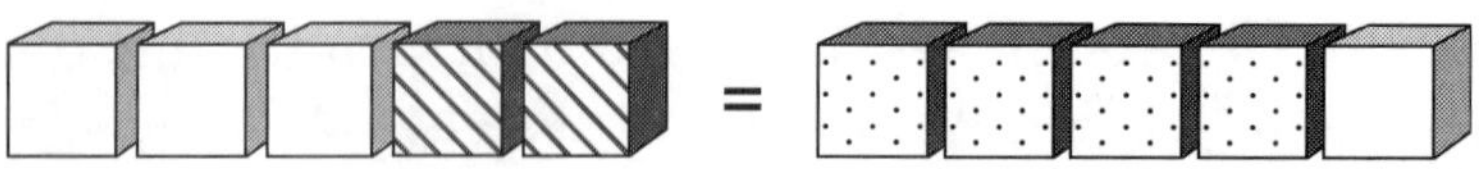

Lesson 4 — Charitable Donations

Example Santos earns $21,400 a year in his business. His expenses total about $1,600 per month. In addition, he donates $50 to two different charities twice each year. He can deduct these donations when he calculates his state income tax. What effect do these donations have on the state income taxes he pays?

Solve

Step 1: Calculate how much Santos donates to the two organizations in one year.

$50 × 2 × 2 = $200 He donates to 2 organizations 2 times a year

Step 2: In his state, donations are subtracted from income and not added to expenses. Subtract this amount from his annual income to find the effect.

$21,400 – $200 = $21,200

Answer the Question

Step 3: These donations reduce his taxable income by $200, to $21,200.

Now try these problems.

1. Joei, who lives in the same state as Santos, earns $31,900 a year. Her expenses total about $2,500 per month. She donates $25 to three different charitable organizations each quarter. What effect do these donations have on the taxes she pays?

   A Reduce taxable income by $75

   B Reduce taxable income by $300

   C Increase taxable expenses by $75

   D Increase taxable expenses by $300

2. *TooCoolTots* is a not-for-profit nursery school. The school holds a big yard sale. Eighteen parents donate items for the sale. Half of those parents also donate their time to work at the sale. What percent of the parents who donate items also donate their time?

   Answer: Of the parents who donate items, ______% of them also donate their time.

3. The athletics club holds a swimathon to raise money for a children's hospital. Out of 53 sponsors, 28 pledge 10 cents per lap for each of ten swimmers. The other sponsors pledge 15 cents a lap for each of eight swimmers. Each swimmer swims 30 laps. How much money can the club collect for the charity? Circle the items you will use as you calculate. Then find the answer.

53	(28 × \$0.01)	(28 × \$0.10 × 10)	(25 × \$0.15 × 8)
30	(25 × \$0.025)	18 × (\$0.10 + \$0.25)	(53 × 30 × \$0.35)

Answer: ______________________________ = $________

4. Each day for 120 days before a big kayak race, the organizers auction off a special T-shirt. The minimum bid is \$25. The maximum bid is \$500. The median bid is \$145. Total proceeds for the 120 T-shirts are \$11,400.

a. Mark and label the line to show the low, high, and median bids. Write a label to show the number of bids between the median and the maximum.

____

b. What is the difference between the mean and median of the winning bids?

A \$50 **B** \$117.50 **C** \$330 **D** \$3,000

☆ *Challenge Problem*

You may want to talk this one over with a partner.

The company where you work donates 1% of its pre-tax profit to charities. In your opinion, what are the financial advantages and disadvantages of this action?

____

____

____

Lesson 5 — New Business Ventures

Example Wesley is starting a business. He rents a small one-room office, furniture, and a copier. His business expenses are about \$650 per week. His living expenses are about \$350 per week. He charges \$30 per hour for his Web design services. He expects to work 35 hours a week. How much money does Wesley have left over each week?

Solve

Step 1: Calculate Wesley's expected weekly income.
$\$30 \times 35 = \$1{,}050$

Step 2: Subtract his weekly expenses.
$\$1{,}050 - \$650 - \$350 = \$50$

Answer the Question

Step 3: Wesley has \$50 left over each week.

Now try these problems.

1. Allysia rents a small garage to use as a workshop. Her business expenses are about \$350 per week. Her living expenses are about \$275 per week. She charges \$25 per hour for her work. She expects to work 30 hours a week. How much is left over each week? Fill in the blanks.

   Answer: Each week, Allysia has ________ – ________ = ________.

2. Daken needs to buy some equipment for his machine repair business. He understands that a small business can deduct capital expenses of up to \$25,000. He buys new tools (\$7,500), equipment (\$9,250), and new signage (\$1,250) as capital expenses. How much can he deduct for capital expenses?

   A \$0 **C** \$16,750

   B \$9,250 **D** \$18,000

3. Nell pays \$87 per week to advertise her portrait business in the local newspaper. In response to her ads, she gets names of

possible customers as leads. By talking to the leads, she converts some of them to actual customers.

a. Fill in the last two columns for each week.

Week	Leads	Customers	Percent of leads converted to customers	Cost per lead
1	24	6		
2	7	4		
3	15	7		

b. What is the mean cost per lead of her advertising program?

Answer: b. Mean cost per lead = ________.

4. To attract customers, Kent offers weeding services at $15/hour below his cost. Once he has a weeding customer, he hopes to sell that customer landscaping services. Landscaping services cost $45 per hour, which is $10 an hour above his cost. He draws up a chart to show projected profit for three different yards. Check the entries in the profit column. Write in any corrections.

	Weeding services	Landscaping services	Profit
Yard 1	1 hour	3 hours	$ 5
Yard 2	2 hours	5 hours	$30
Yard 3	3 hours	10 hours	$55

☆ *Challenge Problem*

You may want to talk this one over with a partner.

Typical attendance at Monday night movies is 70 people. To increase attendance, Venn advertises that the first 50 people will get a movie poster free. On the next Monday night, the first 50 people make up 62.5% of the total audience. The poster costs Venn $3 each to make. Tickets cost $7 each. Is this promotion a good idea? Why?

____

____

____

____

____

Unit 4 Review

Review What You Learned

In this unit you have used mathematics to solve many problems. You have used mental math and estimation, practiced basic operations, solved equations, and used statistics and probability.

These two pages give you a chance to review the mathematics you used and check your skills.

✔ Check Your Skills

1. In the first three months of the year, Diendro spends $4,200, $5,100, and $2,005. He cannot afford to exceed a total of $36,000 in expenses for the year. What is the mean amount he can spend per month for the rest of the year? Fill in the blanks on his work sheet.

This Year's Expenses

January	$4,200
February	$5,100
March	$2,005
TOTAL (3 months)	$_______
Maximum expense money available for the year	$_______
Remaining expense money for the rest of the year	$_______
Mean expenses per month for the rest of the year	$_______

If you need to review, return to lesson 1 (page 41).

2. Emma reviews her net worth. Net worth is the difference between assets and liabilities. Emma's assets include property, such as jewelry, a car, art supplies, and her bank balance. Her liabilities include financial obligations, such as her art school loan and current bills. Her net worth is $7,400. What is the difference between the value of her assets and her liabilities?

A –$7,400 **C** $7,400

B $0 **D** $14,800

If you need to review, return to lesson 3 (page 45).

3. Sandra earns $22,300 a year. Her expenses total about $1,700 per month. In addition, she donates $35 to two different charities three times a year. These donations can be deducted from her income. What effect do these donations have on the taxes she pays?

Answer: ______________________________

If you need to review, return to lesson 4 (page 47).

4. Melanie works out of her home. Her business expenses are about $300 per week. Her living expenses are about $405 per week. She charges $25 per hour for her catering services. She expects to work 30 hours a week. Write an expression for the amount left over each week.

Answer: ______________________________

If you need to review, return to lesson 5 (page 49).

5. To attract customers, Adalberto offers boat cleaning services at $10/hour below his cost. Once he has a customer, he hopes to sell his boat painting services. Boat painting services cost $45 per hour, which is $15 an hour above his cost. His first customer has a boat that takes 3 hours to clean and 7 hours to paint. Circle the entry below that shows the profit or loss on this boat.

Boat cleaning service	Boat painting service	Profit or loss
– $10	$105	$95 profit
– $30	$105	$75 profit
$30	– $70	– $40 loss

If you need to review, return to lesson 5 (page 49).

Write Your Own Problem ✍

Choose a problem you liked from this unit. Write a similar problem using a situation and related facts from your own life. With a partner, share and solve these problems together. Discuss the mathematics and compare the steps you used. If you need to, rewrite or correct the problems. Write your edited problem and the answer here.

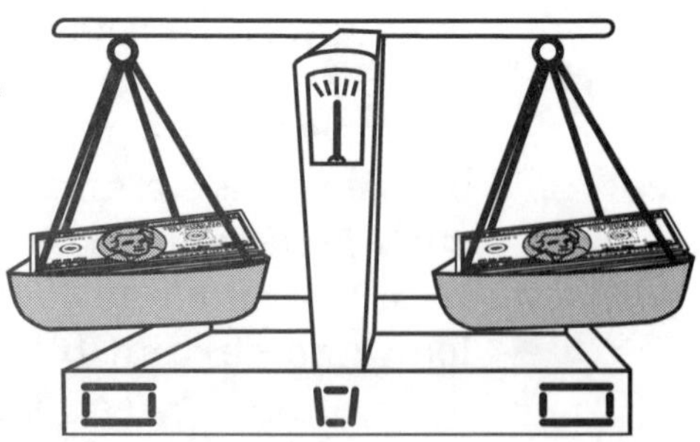

____

____

____

Unit 5 Getting to the Job

Preview

How You Will Use This Unit

Getting to the job involves many things. You may consider different ways to travel and compare the costs and the time it takes. You may consider car pooling or taking the bus or the subway. If you have a car, you think about the cost of keeping it on the road. As you make choices, you will often use math. The math skills you use include mental math and estimation, basic operations and equations, statistics and probability, ratios and proportions, and making measurements (standard and metric).

What You Will Do in This Unit

In this unit, math steps demonstrate how to solve problems. These steps can help you answer questions such as these:

You live an hour from work. Your friend lives 6 minutes from work. You pay $245 rent and $192 for gas each month. Your friend only pays $418. Who pays the least rent plus gas per workday?

When the weather is nice, you row across the bay to work. On other days, you bicycle by the coast road. You can go twice as fast by the road, but it takes the same time. How far is it by the road?

You sell your old car and buy your grandmother's car for $450. The title fee is $12. The fee to transfer the license plate from your old car is $8. How much do you owe?

Your road map shows a scale of 1 centimeter to 25 kilometers. You mark out a route that measures 2.5 centimeters. How many kilometers is this route?

What You Can Learn from This Unit

When you complete this unit, you will have used mathematics to work problems related to getting to the job. These problems are similar to those that may actually occur in your daily life.

Lesson 1 → Cost Compared to Time

Example Meg lives an hour from work. Jaci lives 6 minutes from work. They both work 20 days a month. Each month, Meg pays $245 rent and $192 for gas. (She only uses her car for work.) Each month, Jaci pays $418 rent and nothing for gas. (She walks to and from work.) Meg says her cost per workday is less than Jaci's, even though she has to pay for gas. Jaci says her cost per workday is less than Meg's. Who is correct?

Solve

Step 1: To find Meg's costs, add her rent and cost for gas.

Meg: $245 + $192 = $437

Step 2: Jaci's costs are only her rent.

Jaci: $418

Step 3: To find the costs per workday, divide each total by 20.

Meg: $437 ÷ 20 = $21.85
Jaci: $418 ÷ 20 = $20.90

Answer the Question

Step 4: Jaci is correct.

Now try these problems.

1. Jared lives an hour from work and works 20 days a month. Lamar lives 6 minutes from work and works 24 days a month. Jared pays $205 rent and $110 for gas each month. (He only uses his car for work.) Lamar pays $395 rent and nothing for gas each month. (He walks to work.) Who pays the lesser amount for rent and transportation per workday?

   Answer: ______________________________

2. Tiah works nine miles from home. She drives to and from work 5 days a week. She drives 30 mph. Her car gets 20 miles per gallon. Gas costs her $1.37 per gallon. She changes to a four-day work week schedule. Circle the correct row to show the answers

to the following questions. How many miles of driving does she save? How much travel time does she save? How much gas money does she save?

Miles saved per week	Travel time saved per week	Gas money saved per week
9 miles	18 minutes	\$0.62
18 miles	36 minutes	\$1.23
72 miles	1 hour 12 minutes	\$1.85

3. The ferry costs \$3 a trip. But the waiting in the ferry line takes 45 minutes. Frederico can take a different route. This other route goes south for 15 miles, crosses a short bridge over the river, and comes back north 15 miles. His truck gets 10 miles to the gallon. He pays \$1.45 per gallon. He drives 55 mph. Draw lines to link the matching items. Which route is time effective? Which route is cost-effective?

Ferry	**Time effective**
Road & Bridge	**Cost-effective**

4. Morgan lives 15 miles from work. Today, he drives 60 mph all the way. (Normally he is not in a hurry, and drives 50 mph all the way.) Just as he drives into the parking lot, he gets stopped for a speeding ticket. The ticket costs him \$75, and six minutes of his time. How much earlier or later does he get to work today? How much does each minute driving at 60 mph cost him?

Answer: Morgan gets to work ______ minutes __________ (earlier/later).

Each minute cost him ________.

☆ *Challenge Problem*

You may want to talk this one over with a partner.

You work in the city. Each day you leave your car at the train station (where parking is free to commuters) and take the train into the city. The train journey takes 1 hour. You read on your way down and back. What do you need to know to decide if traveling by train is cost-effective or time effective for you?

____

____

Lesson 2 — Considering Alternatives

Example When the weather is nice, Erin rows across the water to the *Aquarium-on-the-Bay,* where she works. The distance across the bay by boat is one mile. On other days, she bicycles along the coast road. Her speed by bicycle is triple her speed by boat. But it takes her the same time to travel to work either way. What is the distance along the coast road?

Solve

Step 1: Underline the sentence that tells you how fast Erin can travel by bicycle.

<u>Her speed by bicycle is triple her speed by boat.</u>

Step 2: Use *s* for her speed by boat, in miles per hour (mph). Use this variable, *s*, to write a math expression giving Erin's speed by bicycle.

Erin's speed by bicycle ⇒ $3s$.

Recall that distance = speed × time. Use *d* for the distance along the coast road.

Write an equation that gives Erin's time by boat.

Write a second equation for her time by bicycle.

$t_{\text{boat}} = 1 \div s$	Time = distance ÷ speed
$t_{\text{bicycle}} = d \div 3s$	Notice that the distance is in miles.

Step 3: Read the problem again. Underline the sentence that tells you the time it takes Erin by bicycle and by boat. Use this information to combine the two equations for time. Solve for *d*.

<u>But it takes her the same length of time to get to work.</u>

$1 \div s = d \div 3s$	Multiply both sides by $3s$ (the common denominator).
$3 = d$	

Answer the Question

Step 4: The distance along the coast road is 3 miles.

✏ Now try these problems.

1. When it snows, Ulrich skis to work. The distance is 2.5 miles in a straight line across the park. Otherwise, he skateboards along the streets. Both routes take him the same time. His speed on his skateboard is $\frac{7}{5}$ of his speed on skis. But it takes him the same length of time to get to work. What is the distance by the route along the streets?

   Answer: The distance by the street route is __________.

2. Nick skates to work. The journey is 5 miles. He skates at 7.5 miles per hour. The trolley travels twice as fast as Nick. But it makes three stops, each four minutes long, on the way. Does Nick beat the trolley, or does the trolley beat Nick? Use the timeline to show what fraction of an hour each trip takes. Circle the winner.

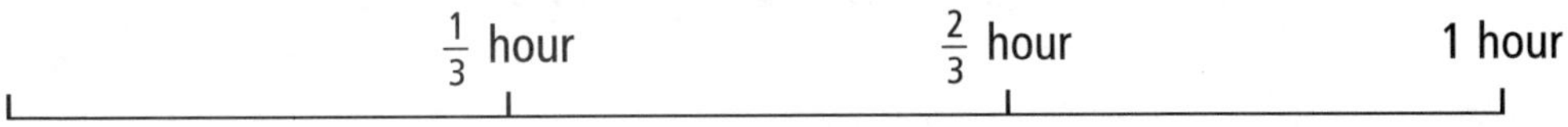

3. Milan buys a van for $6,500. He puts down a $500 deposit, and pays $125 per month. He also pays $110 per month, on the average, for gas. He charges four coworkers $30 a month each to carpool to work. How much does he have to pay over four years in addition to the amount his riders pay him?

   A $740 **B** $5,520 **C** $6,020 **D** $11,780

4. Daria leaves her car in the parking garage at 6:45 a.m., and takes the bus into town. She picks her car up at 6:15 p.m. Circle the smallest bill she can hand to the attendant to cover the parking fee.

Parking garage fees
$1.50 for first $\frac{1}{2}$ hour.
$2 for the next hour
$1.00 for each of the next three hours.
$0.50 for the next 19 hours (or partial hours).

☆ *Challenge Problem*

You may want to talk this one over with a partner.

You share a small plane ride from home to a conference. It takes $\frac{1}{2}$ hour to get to the airport, another $\frac{1}{4}$ hour waiting for takeoff, and $\frac{1}{2}$ hour on arrival to park and tie down the plane. Then the bus ride to the conference takes 15 minutes. The plane flies at twice the speed of the car on the same trip. How long must the car trip take before it is quicker to go by air? Explain your answer.

____

____

Lesson 3

Staying on the Road

Example Michie buys her grandmother's car for $450. The title fee is $12. (If she waits more than 15 days, she must also pay a $15 late fee.) She pays 6% sales tax on the price of the car if she does *not* buy it from a direct relative. The fee to transfer the license plate from her old car to her grandmother's car is $8. Sixteen days after buying the car, she pays the fees. How much does she owe in tax and fees?

Solve

Step 1: Underline the items about fees and taxes.

The title fee is $12. ...a $15 late fee ... 6% sales tax.
The fee to transfer the license plate ...is $8.

Step 2: List the money that she owes for each item. Explain any late fees or exemptions.

Title fee = $12 + $15	Title fee includes late fee because she is 16 days late.
Sales tax = $0	No sales tax because she buys the car from a direct relative.

Fee to transfer the license plate = $8

Step 3: Use mental math to find the total.

$12 + $15 + $8 = $35

Answer the Question

Step 4: Michie owes $35 in tax and fees.

Now try these problems.

1. Trent buys his brother's car for $1,000. The title fee is $12. (After 15 days, he must also pay a $15 late fee.) He pays 6% sales tax on the price of the car if he does *not* buy it from a direct relative. The fee to transfer the license plate from his old car to his brother's car is $8. Fourteen days after buying the car, he pays the fees. How much does he pay?

A $20 **B** $35 **C** $80 **D** $95

2. Inspection stickers cost $10 and are good for one year. Brittany-Lyn takes her motorcycle to an official testing station on September 4th. The headlight is defective. She has 10 days to get it fixed without an extra $10 charge. She goes back in exactly two weeks. Circle the bill that exactly covers her charge.

3. On September 30, 2002, Cass buys commemorative plates for his passenger car. In April 2003, he buys a truck and a trailer. What are his vehicle licensing fees? Complete the entries in the table.

Vehicle	Fee if paid by Oct. 1, 2002	Fee if paid after Oct. 1, 2002	Fee if purchased after March 31, 2003	TOTAL
Passenger car	$35	$45	$17.50	
Commemorative plates	$65	$75	$32.50	
Truck	$35	$45	$17.50	
Trailer	$12	$22	$6	
ANNUAL FEES				

4. The yearly finance charge on a used car is 10% of the original unpaid balance. Jacob buys a used car for $4,500. He puts $700 down as a deposit. He agrees to pay the balance off in three years. What is his total finance charge?

Answer: ___________

☆ *Challenge Problem*

You may want to talk this one over with a partner.

You buy a used car from your father for $3,200. You pay a deposit of $800, and agree to pay him the rest in equal installments over the next year. You pay a $12 title fee right away, as well as an $8 license plate transfer fee, and a $10 fee for an inspection sticker. What is the mean amount per month (less the deposit) that the car has cost you over the year? Explain your answer.

____

____

____

Lesson 4 — Avoiding Traffic

Example Yosi's road map gives a scale of 1 centimeter to 15 kilometers. He marks out a route to avoid Main Street traffic. The route measures 2.5 centimeters on the map. How many kilometers long is his route?

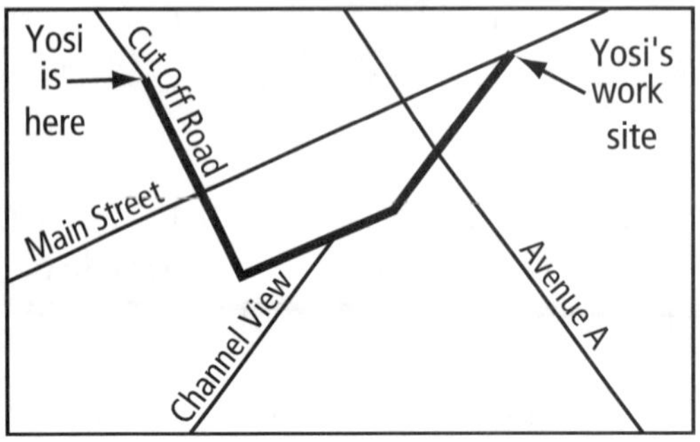

Solve

Step 1: Underline the sentence that tells you the scale for the road map.

Yosi's road map gives a scale of 1 centimeter to 15 kilometers.

Step 2: Write the ratio given by the scale.

$$\frac{1 \text{ centimeter}}{15 \text{ kilometers}}$$

Step 3: Use d for the distance in kilometers of Yosi's route. Write a ratio of the distance on the map to the actual distance in kilometers of Yosi's route.

$$\frac{2.5 \text{ centimeters}}{d \text{ kilometers}}$$

Step 4: Write a proportion with these two equal ratios. Solve for d.

$\frac{1}{15} = \frac{2.5}{d}$ Notice that both ratios have centimeters on top.

$d = 2.5 \times 15$ Cross multiply to find d.

$d = 37.5$ kilometers

Answer the Question

Step 5: Yosi's route is 37.5 kilometers long.

Now try these problems.

1. MaryAnne's road map has a scale of 1 centimeter to 50 kilometers. She marks out a route to avoid highway construction

work. Her route measures 3 centimeters on the map. Write a proportion to find d, the distance in kilometers of her actual route.

Answer: ________________________________

2. In the mornings, Tandy drives through town on Main Street to avoid freeway traffic. Main Street is a diameter of the circular freeway route. The distance from A to B on the freeway is 31.42 kilometers. How many kilometers is his journey across town?

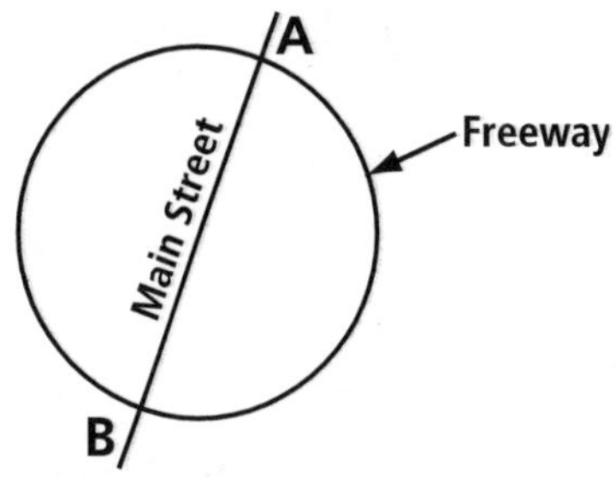

   A 3.142 kilometers
   B 10 kilometers
   C 15.72 kilometers
   D 20 kilometers

3. Monica lives 50 miles from the company where she works. When she leaves home before 6.45 a.m., she clocks in at work at 7:30 a.m. When she leaves home after 7 a.m., it takes her twice as long to get to work. What time does she clock in when she leaves home at 7:10 a.m.? Mark that time on the clock.

4. To avoid city traffic, Julio takes the freeway. The freeway is often crowded, but he can usually drive at 90 kilometers per hour. His reaction time behind the wheel, from seeing a danger to hitting the brake pedal, is $\frac{1}{2}$ second. How many meters will he travel in that $\frac{1}{2}$ second?

Answer: __________

☆ *Challenge Problem*

You may want to talk this one over with a partner.

How many possible routes are there to get from home to work? You are equally likely to take any of the routes. What is the probability that you take routes 1 and 3?

____

____

____

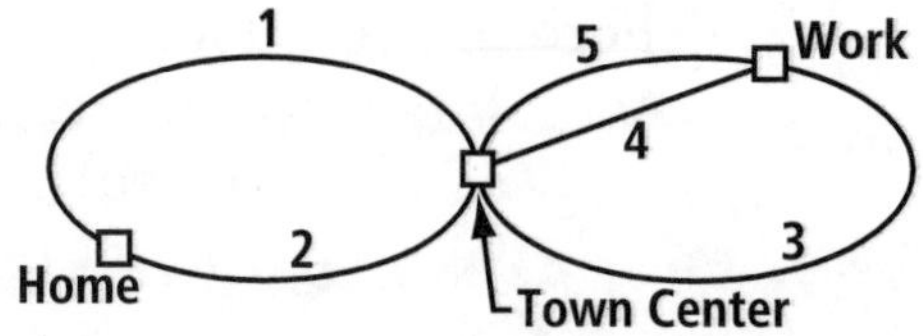

Lesson 5 Marking Routes

A Graphic Game (for Two or More Players)

The goal of this game is to draw arcs. Each person builds on what the previous person has drawn.

Materials

Large sheet of paper, colored pencils

Directions

1. Choose a different colored pencil for each player. Sit with players around the sheet of paper.
2. Player 1 marks three dots on the sheet of paper.
3. Player 2 then draws an arc, using these rules:
   a. You can draw an arc that connects two dots.
   b. You can only draw an arc that does *not* cross another arc.
   c. You can draw an arc only when the first (beginning) dot and the second (end) dot both already have no more than three lines touching them. This means that any dot can have at most four lines coming from it.
   d. After you draw the new arc, mark a new dot somewhere on it.
4. Players alternate turns, each drawing an arc and a new dot on it.
5. The winner is the last player to draw an arc that follows the rules.

Before you play the game, try these warm-up problems.

1. You play *Marking Routes*. There are four dots on the sheet of paper. How many arcs are there? Explain your answer.

   Answer: ____________________

   ____

2. There are ten dots on the sheet of paper. All but two dots are the end points of three arcs. Those two dots, A and B, are the end points of two arcs each. What is a possible next move?

   A Join dot A to a new dot.
   B Join dot A to itself.
   C Join dots A and B if the arc does not cross another arc.
   D Join dot B to one of the other eight dots.

Sample game of *Marking Routes*

Start

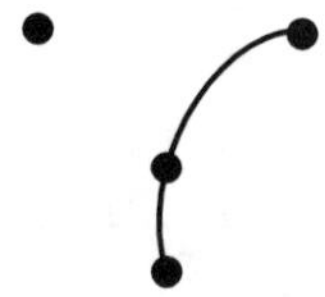
Move 1

Move 2

Move 3

Move 4

Move 5

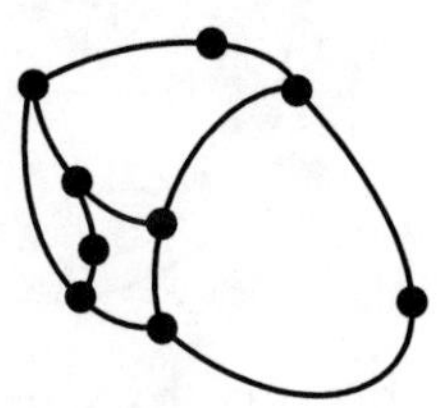
Move 6

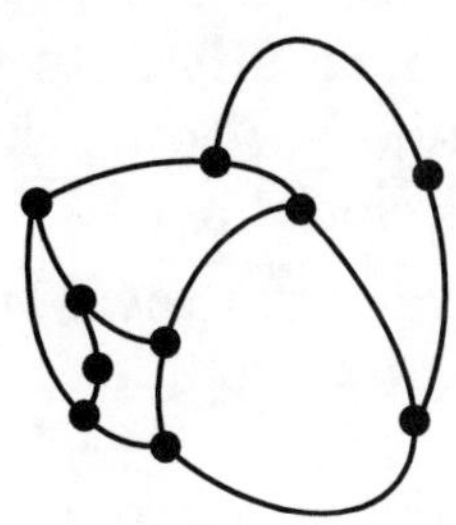
Move 7

Unit 5 Review

Review What You Learned

In this unit you have used mathematics to solve many problems. You have used mental math and estimation, statistics, probability, ratios and proportions, practiced basic operations, solved equations, and made measurements (standard and metric).

These two pages give you a chance to review the mathematics you used and check your skills.

✔ Check Your Skills

1. Charlie lives 30 minutes from work. Trey lives 10 minutes from work. They both work 22 days a month. Charlie pays $260 rent, and $205 for gas each month to get to work. Trey pays $478 rent and nothing for gas each month. (He cycles to work.) Who pays the least rent plus gas per workday?

   Answer: ______________________________

   If you need to review, return to lesson 1 (page 54).

2. Aisha lives 24 miles from work. Today, she drives 45 mph all the way. (Normally she is in a hurry, and drives 60 mph all the way.) The clock shows the time at which she normally gets to work. How much later does she get to work today? Mark that time on the clock face.

   If you need to review, return to lesson 1 (page 54).

3. When the weather is nice, Roame sails across the bay to work. The distance is two miles. On other days, she runs along the coast road. She can run twice as fast as she can sail. But it takes her as long to get to work. What is the distance along the coast road?

   Answer: ______________________________

   If you need to review, return to lesson 2 (page 56).

4. Fran buys a friend's car for $150. The title fee is $12. (If she waits more than 15 days, she must also pay a $15 late fee.) She pays 6% sales tax on the price of the car if she does *not* buy it from a direct relative. The fee to transfer the license plate is $8. Three weeks after buying the car, she pays the fees. Circle the

items for which she owes money. Write the amount she owes under each item. What is the total amount that she owes?

Title Fee	Sales tax	Fee to transfer license plate	Late fees
________	________	________	________

Answer: Fran owes ________.

If you need to review, return to lesson 3 (page 58).

5. Your street map shows a scale of 1 centimeter to 5 kilometers. On the map, the shortest distance to the job site measures 6.5 centimeters. How many kilometers is it to the job site?

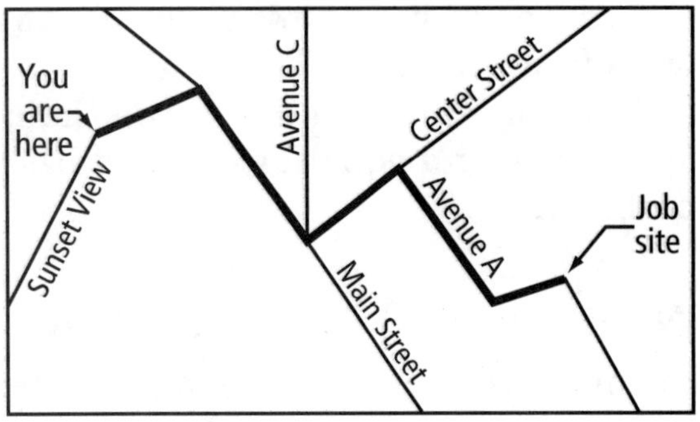

Answer: ________________

If you need to review, return to lesson 4 (page 60).

Write Your Own Problem

Choose a problem you liked from this unit. Write a similar problem using a situation and related facts from your own life. With a partner, share and solve these problems together. Discuss the mathematics and compare the steps you used. If you need to, rewrite or correct the problems. Write your edited problem and the answer here.

____

____

____

____

Unit 6 Getting Paid

Preview

How You Will Use This Unit

Getting paid involves many things. You may get paid hourly, or earn fees per project, or receive a salary. You may compare your gross pay and net pay, and calculate payments such as taxes and insurance. As you think about getting paid, you will often use math. The math skills you use include mental math and estimation, basic operations and equations, and statistics and probability.

What You Will Do in This Unit

In this unit, math steps demonstrate how to solve problems. These steps can help you answer questions such as these:

Your boss gives you a pay scale table. The table shows your annual gross and net pay, and a possible increase in hourly rate in six months. What is your net pay after the increase?

You make $12.50 per hour on regular time. On overtime, you make time-and-a-half. This week, you work 43 hours. How much money do you make this week?

You sign up for a bill paying service. One option offers 12 bills paid for $5.98. The other option offers $0.98 per bill. What number of bill payments makes each option more cost-effective?

This year, your taxable income is $19,500. This puts you in the 15% tax bracket. How much income tax do you pay?

What You Can Learn from This Unit

When you complete this unit, you will have used mathematics to work problems related to getting paid. These problems are similar to those that may actually occur in your daily life.

Lesson 1

Gross or Net?

Example Scott's boss gives him a table that shows the company pay scale for a 40-hour week. He tells Scott that 30% of the gross pay is withheld for taxes and insurance. Scott wants to see the amounts if he gets the same raise again at the end of the year. For these amounts, complete the last line of the table. (Round all entries, except for hourly rate, to the nearest dollar.)

	Gross Annual Salary	Hourly Rate	Net annual pay after 30% withheld	Net pay for each of 26 pay periods
Now	$21,000	$10.10	$14,700	$565
6-month raise	$22,000	$10.58	$15,400	$592
12-month raise	$______	$_____	$______	$_____

Solve

Step 1: Find the amount of the 6-month pay increase. To find the 12-month hourly rate, add that same amount.

$10.58 – $10.10 = $0.48
6-month pay increase for hourly rate

$10.58 + $0.48 = $11.06
hourly rate after 12-month raise

Step 2: Next, calculate Scott's gross annual salary at this new rate.

$11.06 × 40 × 52 ≈ $23,005
Multiply rate by hours/week and weeks/year.

Step 3: Find the net annual pay after 30% is withheld. First, write the percent of annual salary that he takes home. Then, calculate that percent of gross annual salary.

100% – 30% = 70%
70% of $23,005 = 0.7 × $23,005
≈ $16,104

Step 4: To find the net pay for each of 26 pay periods, divide net pay by number of pay periods in one year.

$16,004 ÷ 26 ≈ $619

Answer the Question

Step 5:

Second raise	$23,005	$11.06	$16,104	$619

✏ Now try these problems.

1. Use the table in the example. Scott received both the 6-month raise and 12-month raise. He works out his gross and net pay for a *third* raise at 18 months of the same amount. Fill in the table entries for this. Round entries, except for hourly rate, to the nearest dollar.

Third raise				

2. Hermani's gross income is $400 per week. The payroll office withholds $60 for federal income tax, one third of that for state income tax, and $24 for social security tax. What is Hermani's net pay per week?

   A $104 **B** $183 **C** $296 **D** $340

3. Nan makes $12 per hour. She works 40 hours per week. Payroll subtracts 15% for federal income tax, 5% for state income tax, 6.1% for social security tax, and 1% for union dues. What is the dollar amount of these items: her gross pay, federal income tax, state income tax, union dues, and net pay per week? Draw lines to connect each item with the correct amounts.

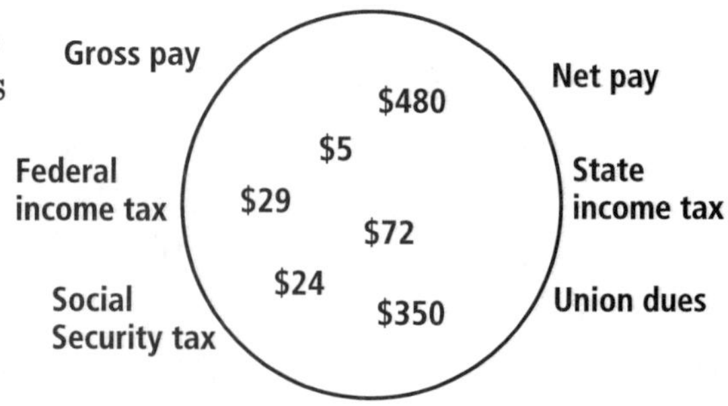

4. Clyde makes $8,000 as a bonus at the end of the year. Clyde invests 25% of this amount in an account that earns him 5% interest annually.

   a. How much does his investment earn him in one full year?

   b. Does his investment plus interest equal 30% of the $8,000? If not, how much more or less is it?

   Answer: a. ______ **b.** ________________________________

☆ *Challenge Problem*

You may want to talk this one over with a partner.

Your bank looks at your gross income of $35,000 to qualify you for a house mortgage. Your net annual income is 70% of your gross annual income. The bank estimates 29% of your gross income as the maximum amount of your house payments. In your opinion, is this realistic? Explain your answer.

____

____

Lesson 2 — Points & Numbers

A Game of Numbers (for Two or More Players)

The goal of this game is to earn points by using facts about a number. First, you pick up a card. Then, you assign points to the number on that card, using the rules on the scorecard.

Materials

Game cards and a copy of the scorecard for each player (on the next page), sheets of paper for each player (or a pad), pencils, a timer, one large piece of paper for the scorekeeper

Directions

1. Each player has several sheets of paper and a pencil. Shuffle the game cards and place the deck face down in the center. Sit with players around the play area. One player acts as scorekeeper. Players decide at the start how many minutes to allow per round.
2. Player 1 draws a game card and places the card, face up, on the discard pile. This player then sets the timer for the round. Each player, including player 1, uses the score card to assign points to the number. Players write a list on one piece of their paper. They list the letter of the rule and the points assigned to the number by that rule. When the timer goes off, they stop writing and total the points they have assigned.
3. When all players have written their lists and totals, they pass them to the scorekeeper. Any player can contest an item on any list, in which case the player who wrote it must defend or change it. The scorekeeper writes on a score sheet the total of the round for each player.
4. Each player takes a turn drawing a card and setting the timer. When a player draws a *create-a-number* card, if that player can produce the number, that player earns 25 bonus points.
5. The game ends when all cards have been drawn. The winner is the player with the greatest point score.

Before you play the game, try this warm-up problem.

1. You draw an *assign-points-to-a-number* card that has the number 182. What is the total number of points you can assign to this number?

   A 30 **B** 31 **C** 41 **D** 46

Cards for *Points & Numbers*

Scorecard

A. 1 point if the number is an odd number
B. 2 points if the number is an even number
C. 3 points if the number is a multiple of 3 or 5
D. 4 points if the number is a multiply of 7
E. 5 points if the sum of the tens digit and the hundreds digit is greater than 6
F. 5 points if the product of the hundreds digit and the thousands digit is greater than 40 (may not apply)
G. 10 points if the difference between the thousands digit and the ten-thousands digit is less than 3 (may not apply)
H. 10 points if a digit appears two or more times
I. 20 points if the sum of all the digits is less than 20

Game cards

Assign-points-to-a-number cards:

1. Make at least 25 cards, writing on each a different 3-, 4-, or 5-digit number.

Create-a-number cards:

2. Create 6 special cards with these directions on them.

Create a number that is less than 1,493 and fits rules B and D (but may fit others, too).	**Create a number that is greater than 875 and less than 7,002 and fits rules A, C, and H.**
Create a number that does *not* fit either rule A or rule B.	**Create a number for which at least 4 of the scorecard rules apply.**
Create a number that fits rules G, H, and I.	**Create a number that fits rules A, D, and F.**

Lesson 3 — Pay Alternatives

Example Rindle's regular workweek is 40 hours. He makes $11 per hour on regular time. On overtime, he makes time-and-a-half. This week, he works 47 hours. What is his pay this week?

Solve

Step 1: Calculate the amount that Rindle earns for his regular 40 hours of work.

Pay for regular hours ⇒ 40 × $11 = $440

Step 2: Find the number of overtime hours that Rindle works.

Overtime hours ⇒ 47 hours – 40 hours = 7 hours

Step 3: Next, write an expression for his overtime pay rate.

Rate for overtime hours ⇒ $11 × 1.5
Time-and-a-half means 1.5 times the regular pay rate.

$11 × 1.5 = $16.50

Step 4: Then, multiply the overtime pay rate by the number of overtime hours.

Pay for overtime hours ⇒ $16.50 × 7 = $115.50

Step 5: Add Rindle's regular pay and his overtime pay for the week.

$440 + $115.50 = $555.50

Answer the Question

Step 6: Rindle's pay for the week is $555.50.

✏ Now try these problems.

1. Nina's regular workweek is 40 hours. She makes $12.50 per hour on regular time. On overtime, she makes time-and-a-half. This week, she works 44 hours. She writes an expression for her pay this week. What mistake does she make? Circle and explain her mistake. How much more or less than this amount does she make?

   Gross pay this week = 40 × $12.50 + 4 × ($12.50 + 1.5)
   = $556

   Answer: ____________________

2. Each year, Meredith receives a cost-of-living increase. Over three years, her salary has gone from $12,000 to $12,600 to $13,230.

   a. What is the percent of the cost-of-living increase each time?

   Answer: a.________

   This year, she receives the same cost-of-living increase, plus an 8% merit increase.

   b. What is her merit increase and her new salary for this year? Circle the correct answer.

Merit increase	New salary
$630	$14,288
$1,720	$14,888
$1,058	$14,950

3. *CityCentral* pays Raul and Fil on the first day of each month. The company automatically deposits Raul's check in his bank account on that day. Fil deposits his check himself. On average, he does this on the fifth day of the month. Each account earns interest. About how many days of interest does Fil lose in a year?

   A 48 **B** 60 **C** 73 **D** 91

4. Rudy gets paid per project, usually 30 days after the project is complete. He charges 30% of the total price as an up-front payment before he starts. He charges another 30% when he presents the first version of the project. The rest he charges when the project is complete. He starts a new project. The total price is $4,000. What are the three payments?

   Answer:____

☆ *Challenge Problem*

You may want to talk this one over with a partner.

You find out that one quarter weighs the same as two pennies. You also find out that one pound of quarters is worth $25. How much is one pound of pennies worth?

____

____

____

____

Lesson 4 — Paying Bills

Example Andrea signs up for an online bill paying service. She has two options. What number of bill payments per month makes each option the most cost-effective one?

Option 1

- *3 months free upfront!*
- $5.95 per month for up to 12 bill payments
- $0.50 for each additional bill payment

Option 2

- *Pay as you go!*
- $0.95 per bill payment
- No limit on the number of bill payments per month

Solve

Step 1: How many bill payments at the option 2 price of $0.95 could she make for $5.95? (The monthly fee for 12 payments in Option 1 is $5.95).

$5.95 ÷ $0.95 = 6.26 bill payments

Step 2: Draw a chart to compare the cost per month under each option for both 6 and 7 bill payments.

	Option 1	Option 2
6 bill payments/month	$5.95	$0.95 × 6 = $5.70
7 bill payments/month	$5.95	$0.95 × 7 = $6.65

Answer the Question

Step 3: For 6 or fewer bill payments, option 2 is more cost-effective.

For 7 or more bill payments, option 1 is more cost-effective.

Now try these problems.

Refer to the information in the example as you work the first two problems.

1. Andrea's online bill paying service increases its prices. Under option 1, the price of up to 12 bill payments changes to $8.29. Under option 2, the price per bill payment changes to $1.00. Now what number of bill payments per month makes each option the more cost-effective?

Answer: For ______ or less bill payments, option ______ is more cost-effective.

For ______ or more bill payments, option ______ is more cost-effective.

2. Barden makes 15 bill payments per month. He looks at Andrea's online bill paying service with the recent increases. He sees that there is still a $0.50 charge for each additional bill payment over 12. Which is his more cost-effective option? How much will he pay per month? Circle the correct answer.

   A Option 1, $9.79   **C** Option 2, $13.50

   B Option 1, $15.79   **D** Option 2, $15

3. The credit card company charges Catrione 18.75% annually on unpaid balances. She pays no interest when she pays her balance in full. For the first three months, she pays her balance in full. The fourth month, she pays the minimum of $100 on a balance of $3,054. How much will she owe the fifth month, excluding any new charges. Fill in the blanks to complete her calculation.

New balance	Minimum payment due	Annual interest rate	Total balance owed
$3,054	$100	18.75%	$______ × __________ = $______

4. Tyson's rent bill is $548 per month. A note says, "A 5% late charge is added to payments after the tenth of the month." He usually has $50 left over each month after he pays all his bills. On May 4, he buys a toolset for $21.99. But, he forgets to pay his rent until the fifteenth. How much money does he have left over this month? Circle the exact amount.

☆ *Challenge Problem*

You may want to talk this one over with a partner.

Each small cube is 1 inch on each side. How many small cubes are in the larger cube? How many small cubes do you have to *add* to make the 2 × 2 × 2 cube into a 3 × 3 × 3 cube? How many small cubes do you have to *add* to make the 3 × 3 × 3 cube into a 4 × 4 × 4 cube?

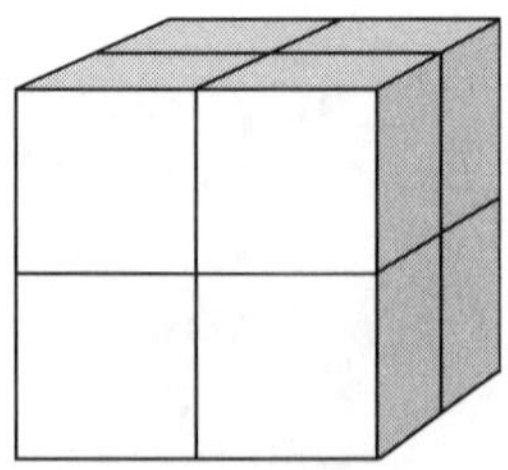

Example The chart shows federal tax brackets for taxable income from $0 to $151,000 and up. Damien uses this chart to estimate his tax. Damien's taxable income is $19,200. What is the income tax on this amount of taxable income?

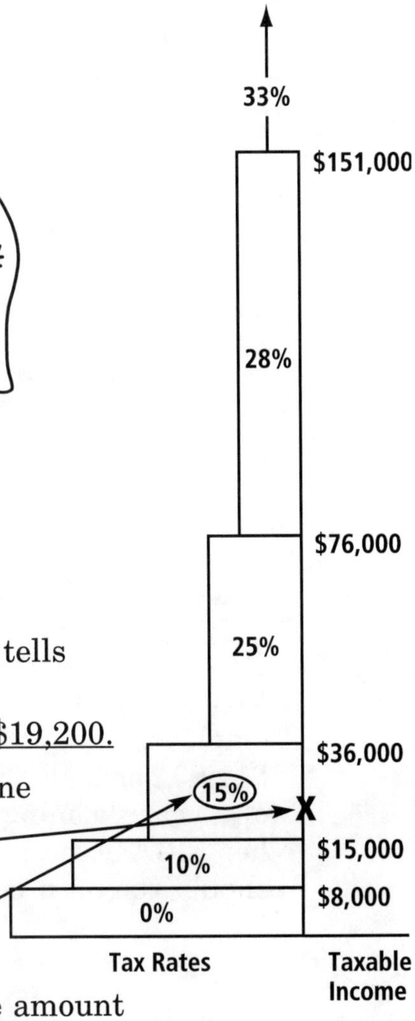

Solve

Step 1: Underline the sentence that tells Damien's taxable income.

Damien's taxable income is $19,200.

Step 2: Make an **X** on the vertical line to show Damien's taxable income.

Step 3: Circle the tax rate for that income bracket.

Step 4: Use the rate to calculate the amount of income tax on that amount.

15% of $19,200 = 0.15 × $19,200
= $2,880

Answer the Question

Step 5: The amount of income tax on $19,200 is $2,880.

✏ Now try these problems.

Refer to the information in the example as you work the first two problems.

1. Britta's taxable income is $36,500. What is the rate in her tax bracket? What is the income tax on this amount of taxable income? Draw a line between the rate and the amount of income tax.

Tax bracket	10%	15%	25%	28%
Income tax	$5,475	$9,125	$10,220	$12.045

2. The Crenshaws' taxable income is $36,750. They make a tax deductible donation of $1,000 to a health research organization. This donation reduces their taxable income. What is the difference in the amount of tax with the donation and without?

   A $368 B $1,000 C $3,575 D $3,825

3. Rusty cannot decide which state he would like to move to for a new job. So he closes his eyes and drops a paper clip on a map of the United States. Then he looks at a chart of statistics on income tax for the various states. What is the chance that, of the 50 states, he picks one with no state income tax?

Categories of State Income Tax	Number of States
No state income tax	7
Limited state income tax	2
More than 3 state income tax brackets	27

   Answer: ___________

4. The payroll department deducts 7.65% of Jasmine's income for FICA (Federal Insurance Contributions Act) tax payments. Her employer pays an equal amount on her behalf. Jasmine earns $11.90 per hour for 20 hours each week. She works 45 weeks out of a year. Jasmine writes an expression for the *total* payment into her FICA account in a year. What factor does she miss? What should the total be?

$$0.0765 \times \$11.90 \times 20 \times 45 = \$819.32$$

   Answer: ____

☆ *Challenge Problem*

You may want to talk this one over with a partner.

A car can be one of your biggest business tax deductions. You can either deduct the actual business-related expenses (gas, oil, repairs, insurance). Or you can deduct 36 cents for each business mile that you drive. This year, you keep a careful log of your expenses. Your total expenses are $965. On average, you have driven 235 miles per month. Which way of deducting saves more in taxes? Explain your answer.

____

____

____

Unit 6 Review

Review What You Learned

In this unit you have used mathematics to solve many problems. You have used mental math and estimation, practiced basic operations, solved equations, used statistics and probability, and made measurements using both standard and metric measurements.

These two pages give you a chance to review the mathematics you used and check your skills.

✔ Check Your Skills

1. On his first day of work, Manuel's boss gives him a pay scale table for a 40-hour workweek. He tells Manuel that 30% of his gross pay is withheld for taxes and insurance. The table shows Manuel his annual gross and net pay now and for a possible raise in six months. Manuel works out his gross and net pay for a second raise of the same amount. Complete the last row. Round all entries, except for hourly rate, to the nearest whole dollar.

	Gross Annual Salary	Hourly Rate	Net annual pay after 30% withheld	Net pay for each of 26 pay periods
Now	$25,000	$12.01	$17,500	$673
6-month raise	$26,000	$12.50	$18,200	$700
Second raise				

If you need to review, return to lesson 1 (page 67).

2. Troise's regular workweek is 40 hours. He makes $9 per hour on regular time. On overtime, he makes time-and-a-half. This week, he works 49 hours. How much *more* money does he make this week than during a regular 40-hour week?

A $13.50 **C** $121.50

B $81 **D** $481.50

If you need to review, return to lesson 3 (page 71).

3. Lewise signs up for an online bill paying service. She has two options. What number of bill payments per month makes each option more cost-effective?

Option 1

- *3 months free upfront!*
- $7.34 per month for up to 12 bill payments
- $0.50 for each additional bill payment

Option 2

- *Pay as you go!*
- $0.85 per bill payment
- No limit on the number of bill payments per month

Answer: For ______ or less bill payments, option ______ is more cost-effective.

For ______ or more bill payments, option ______ is more cost-effective.

If you need to review, return to lesson 4 (page 73).

4. This year, Kalani's taxable income is $26,800. What is the tax on that amount?

Taxable Income	$0 to $8,000	$8,001 to $15,000	$15,001 to $36,000	$36,001 to $76,000
Tax Rate	0%	10%	15%	25%

A $1,769.85

B $3,269.95

C $4,020.00

D $5,400.00

If you need to review, return to lesson 5 (page 75).

Write Your Own Problem

Choose a problem you liked from this unit. Write a similar problem using a situation and related facts from your own life. With a partner, share and solve these problems together. Discuss the mathematics and compare the steps you used. If you need to, rewrite or correct the problems. Write your edited problem and the answer here.

____

____

____

____

____

Unit 7 Teamwork

Preview

How You Will Use This Unit

Teamwork involves many things. You may work as part of team on a project. You may solve problems or complete assignments. You may make plans to complete a job more quickly. Teamwork is also important as you work to improve your community. As you work on a team, you will often use math. The math skills you use include mental math and estimation, basic operations and equations, statistics and probability, and measurements (standard and metric).

What You Will Do in This Unit

In this unit, math steps demonstrate how to solve problems. These steps can help you answer questions such as these:

Your school competes in a robotic competition. The school with the least participation per student population is still able to win an award. Is it your school that does this?

You have been reading while your friend drives. You approach an intersection. A large tree obscures the town names on a sign nearby. Where are you?

Out of fifty-four 11th grade students, and forty-two 12th grade students, 30 sign up for volleyball. One sixth of the 11th grade signs up. What fraction of the 12th grade signs up?

You take 4 hours to mow the park grounds. Your coworker takes 30% longer than you to do the same job. You decide to work together. How long does it take you to mow the grounds together?

What You Can Learn from This Unit

When you complete this unit, you will have used mathematics to work problems related to teamwork. These problems are similar to those that may actually occur in your daily life.

Lesson 1

Group Projects

Example Three schools compete in the Three Points *Young Gardeners* competition. The Three Points *Gardeners Club* makes several awards. One award is for the school with the highest percent of the student population participating. Which school wins this award?

	Westbury High	South Central High	River View High
Total school population	214	419	356
Program participants	15	25	24

Solve

Step 1: Calculate the percent of students participating in each school.

For Westbury High.

$\frac{15}{214} \times 100\% = 7.0\%$

For South Central High.

$\frac{25}{419} \times 100\% = 6.0\%$

For River View High.

$\frac{24}{356} \times 100\% = 6.7\%$

Step 2: Compare the percent of the participation for all the schools.

Westbury High has 7% participation, the highest percent of participation.

Answer the Question

Step 3: Westbury High wins the award.

Now try these problems.

Refer to the information in the example as you work the first two problems.

1. Westbury, South Central, and River View High Schools compete for several awards. The Three Points *Gardeners Club* makes an award for the best use of native plants. This award is won by the school with the second greatest participation per student population. Which school wins this award?

   Answer: ______________________ wins the award for the best use of native plants.

2. One of the three schools wins the award for the greatest participation per student population. How many *more* student participants would each of the other two schools need to have won this award? Write an expression for each answer.

South Central High would need ____________________ so they need about ______ more.

River View High would need ____________________ so they need about ______ more.

3. During a rummage sale, Briony and two friends raise $155 on Friday, $384 on Saturday, and $293 on Sunday. What is the mean amount of money per person they raise?

A $224.00 **B** $277.33 **C** $304.50 **D** $832.00

4. For the team project, Clem finds twice as many Internet reference sites as Taylor. Four of the sites they find are the same.

a. Use x for Clem's unique sites. Use y for Taylor's unique sites. Write an equation to describe these facts.

Answer: a. $x + 4 =$ ____________________

b. What are possible values for x and y? Complete the table for y from 1 to 3. Check that the number of sites Clem finds is always twice the number that Taylor finds.

x	*y*	Clem's total sites	Taylor's total sites
	1		
	2		
	3		

☆ *Challenge Problem*

You may want to talk this one over with a partner.

Angela, Emily, and Franelle work on a project together. The one who does the promotion is an only child. Angela, who dates Franelle's brother, takes more time on her part of the project than the one who does promotion. The one who does design has three sisters but no brothers. What part of the project do they each do? Fill in the chart to explain your answer.

	Angela	Emily	Franelle
Design			
Promotion			
Finances			

Lesson 2 — TeamWorks

♠ A Card Game (for Four Players)

The goal of this game is to work with your partner to win the most points. First, you organize your cards into sets. Then you pass unwanted cards to your partner or around the table.

Materials

Standard card deck

Directions

1. Shuffle, then deal out all the cards. Sit with players around the table, partners opposite each other. One pair of partners plays the other pair of partners.

2. A set is made up of a sequence of three or more cards in the *same* suit (for example, 9, 10, and Jack; King, Ace, 2). Players organize the cards in their hands into sets. Then players state aloud their strong suit. (A strong suit is the suit in which a player thinks a player can win the most sets.)

3. Players then exchange with *their partners* the same number of cards (one or more) that they cannot use. They may also choose to exchange cards that may be helpful to their partners. Next, each player passes one card to the *player on the left* to get rid of unwanted cards.

4. At any time, players can declare a set. To do this, a player places the cards in the set, face up, on the table. (That set cannot be added to later in the game.) Players continue to exchange cards, first with partners, then around the table.

5. The game is over when a player has no more cards. Sets of 3 cards win 1 point. Sets of 4 cards win 2 points. Sets of 5 cards win 3 points, and so on. The winning team is the pair of partners with the most points.

Before you play the game, try this warm-up problem.

1. In your hand, you have the ace, 3, 7, and king of hearts, the 4 of diamonds, and the 5 and 6 of spades. You have declared hearts, your partner has declared spades, and one opponent on the left has declared diamonds. What do you do? Why?

   Answer: ______________________________

Card Sets for *TeamWorks*
Who wins?

Count the points and tell which team of partners wins.

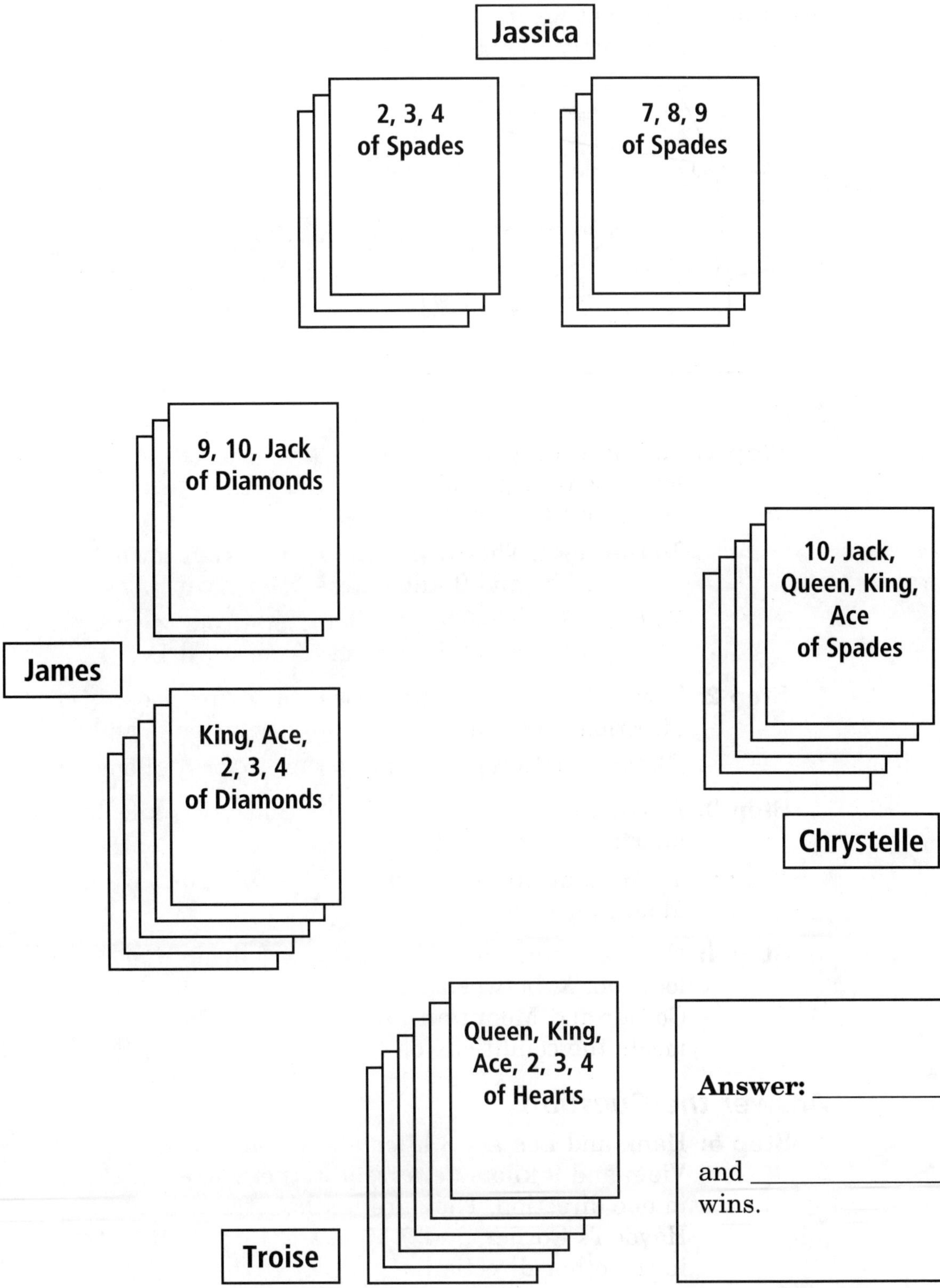

Answer: ____________

and ____________ wins.

Lesson 3 — Problem Solving

Example Hank, Lee's passenger, wakes up just as Lee drives into a gas station. A large tree obscures the town names on a sign nearby. The map gives Hank the general idea of where they are. But he does not want to ask. He tries to figure out for himself where exactly they are. Where are they?

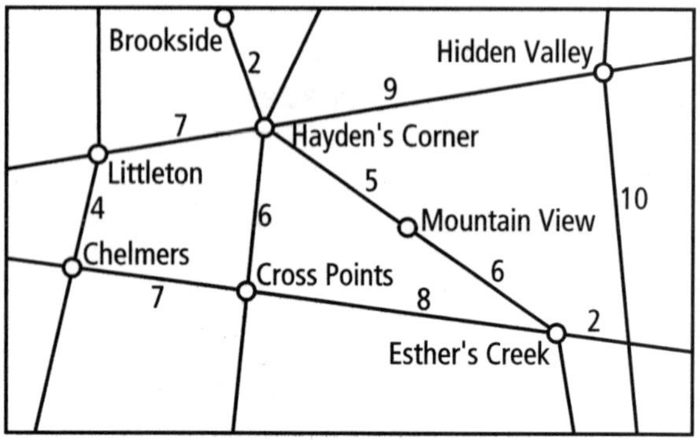

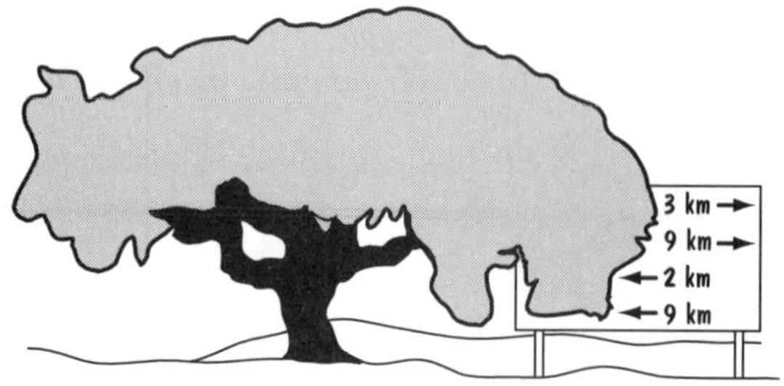

Solve

Step 1: To find where Hank and Lee are, first letter the four towns on the sign, and write the distances Hank and Lee are from these towns.

To the right: They are 3 kilometers from town A, and 9 kilometers from town B.

To the left: They are 2 kilometers from town C, and 9 kilometers from town D.

Step 2: Towns A and C are the closest towns in either direction. Calculate the distance between A and C.

Distance between A and C = (3 + 2), or 5 kilometers.

Step 3: Look on the map to find towns that are 5 kilometers apart.

Towns that are 5 km apart: Hayden's Corner and Mountain View.

Step 4: Now, test and check. Mark the location, **X**, between Hayden's Corner and Mountain View that meets the conditions on the sign.

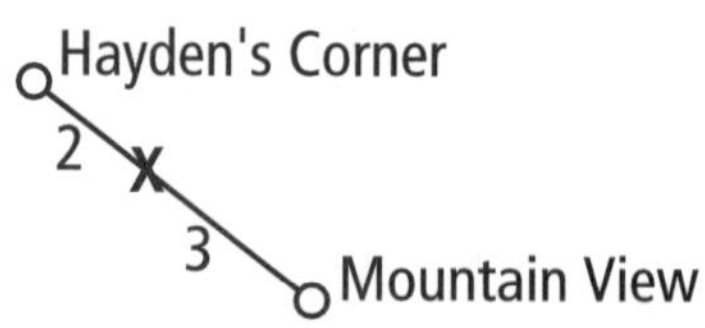

Answer the Question

Step 5: Hank and Lee are 3 kilometers from Mountain View, and 9 kilometers from Esther's Creek, in one direction. They are 2 kilometers from Hayden's Corner, and 9 kilometers from Littleton in the other direction.

✏ Now try these problems.

Use Hank and Lee's map as you work the first two problems.

1. Amber is 2 kilometers from town J in one direction. She is 5 kilometers from town K and 13 kilometers from town L in the other direction. Where is she?

   Answer: Amber is on the road between ________________

   and ________________.

2. You drive from Littleton to Esther's Creek. What is the shortest distance? Circle the towns you go through on the way.

   Answer: The shortest distance is _____ km.

Chelmers   Cross Points   Hayden's Corner   Hidden Valley   Mountain View

3. To settle a disagreement, James and Vijay play 30 rounds of Rock, paper, scissors. In the game, rock breaks scissors, so rock wins; scissors cut paper, so scissors win; paper covers rock, so paper wins. In one round, how many different possible pairs of outcomes are there?

   A 2   **B** 3   **C** 6   **D** 9

☆ *Challenge Problem*

You may want to talk this one over with a partner.

Look at the drawing. What numbers go in the blank spaces?
Hint: How many shapes does the circle labeled 6 touch? Two shapes touch if their perimeters share one or more points. What rule did you use?

A = ________
B = ________
C = ________
D = ________
E = ________
F = ________

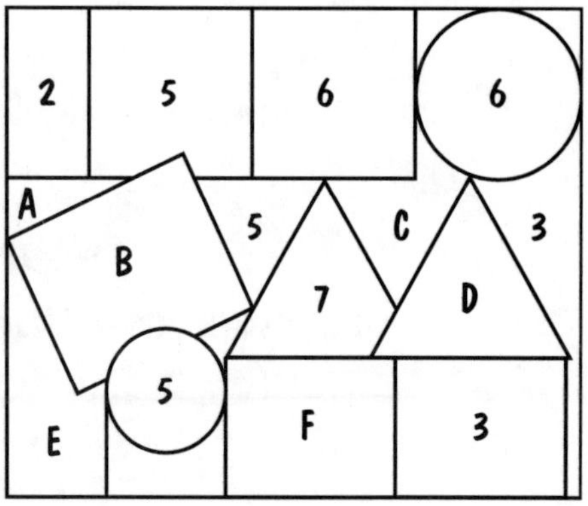

Rule: ____

____

Lesson 4

Building Community

Example There are seventy-five students in the 11th grade and sixty students in the 12th grade. Of these, 21 sign up for volleyball. One fifth of the 11th grade signs up. What fraction of the 12th grade signs up?

Solve

Step 1: Find the words that tell you how many students are in each grade.

11th grade ⇒ 75
12th grade ⇒ 60

Step 2: Underline the words that tell what fraction of the 11th grade signs up.

<u>One fifth of the 11th grade signs up.</u>

Step 3: Now, calculate one fifth of the students in the 11th grade.

$\frac{1}{5}$ of 75 = 15

The word "of" in math usually means to multiply.

Step 4: Next, subtract this number of students from the total that signs up.

21 – 15 = 6

So 6 12th grade students signed up.

Step 5: Now, find what fraction this number is of all 12th grade students.

$\frac{6}{60} = \frac{1}{10}$

Answer the Question

Step 6: One tenth of the 12th grade signs up for volleyball.

✏ Now try these problems.

1. Out of forty people in the sales department, and twenty-two in the advertising department, 21 sign up for water polo. One quarter of the sales department signs up. What fraction of the advertising department signs up?

Answer: ____________ of the advertising department signs up for water polo.

2. In preparation for the fund-raising road rally, 237 students come to the auditorium. The auditorium has 14 rows of seats. There are 16 seats in each row. How many students will *not* get seats?

A 13 **B** 14 **C** 16 **D** 21

3. In the morning, Anjue catches the elevator to the floor where she works. To meet with her project team, she rides the elevator down seven floors. One of her friends works three floors above her floor. After the meeting, Anjue rides up to that floor. Together they ride the elevator halfway to the street level and stop on the 12th floor for lunch. Mark the different levels on the line diagram. On what floor does she meet her project team?

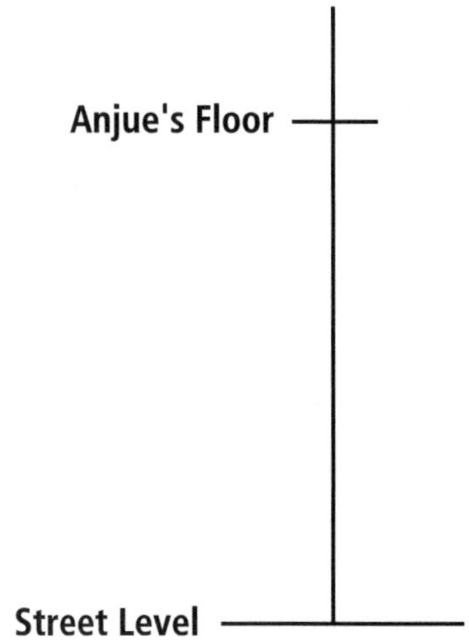

Answer: Anjue meets her project team on

the ________ floor.

4. To hang the art work, Sam and Joe each take a bag of hooks from a box of 12 bags. Sam wonders how many hooks are in each bag. So he counts all the bags and makes a list: 50, 48, 48, 49, 52, 50, 52, 53, 51, 50, 49, 49.
   a. What is the mean number of hooks in the 12 bags (to the nearest whole number)?
   b. Sam picks a bag with exactly 50 hooks. What is the probability that Joe then randomly picks another bag with exactly 50 hooks?

Answer: a. ___________ **b.** ___________

☆ *Challenge Problem*

You may want to talk this one over with a partner.

The local *Teen Leader*s group holds a benefit event for a young student who needs an operation. They set up a dart board and challenge the community. What is the highest score that can be made with 3 darts? What scores, between 1 and the highest, *cannot* be made with 3 darts? (Not all darts hit the board!)

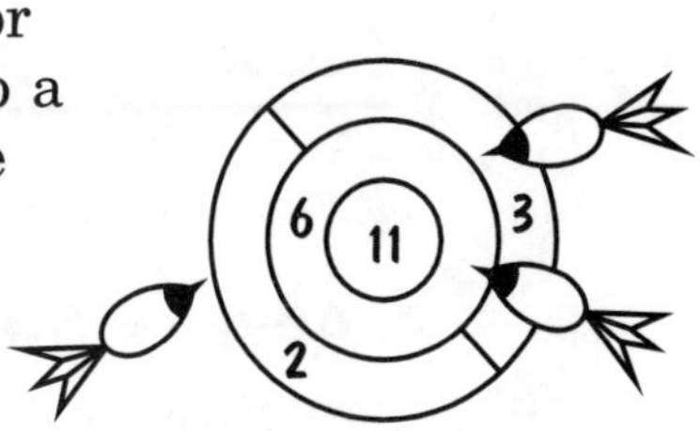

____

____

____

Lesson 5

Collaboration

Example Stephanie takes 2 hours to weed the library gardens. It takes Vy 50% longer than Stephanie to do the same job. How long does it take to weed the whole garden when they work together?

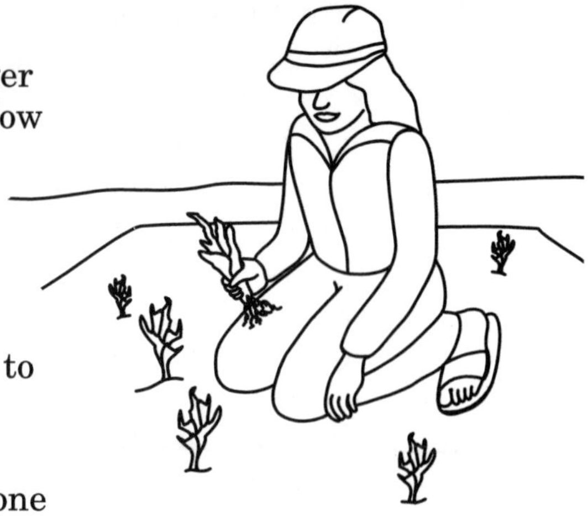

Solve

Step 1: Begin by finding out how long Vy alone would take to weed the whole garden.

2 hours × (100% + 50%)

2(1.5) = 3 hours for Vy alone

Step 2: Now, write expressions for the fraction of the garden each person weeds in one hour.

$\frac{1}{2}$ fraction of the garden weeded in one hour by Stephanie

$\frac{1}{3}$ fraction of the garden weeded in one hour by Vy

Step 3: Together, they can weed the entire garden in *h* hours. Find how what fraction of the garden each person weeds in *h* hours.

$(h)(\frac{1}{2})$ part weeded in *h* hours by Stephanie

$(h)(\frac{1}{3})$ part weeded in *h* hours by Vy

Step 4: The sum of these fractions is one, the entire garden.

$(h)(\frac{1}{2}) + (h)(\frac{1}{3}) = 1$

$h(\frac{1}{2} + \frac{1}{3}) = 1$

$h(\frac{5}{6}) = 1$

$h = \frac{6}{5}$

Answer the Question

Step 5: Together, they can weed the whole garden in $\frac{6}{5}$ hour, or 1 hour and 12 minutes.

Now try these problems.

1. Alex takes 4 hours to mow the park grounds. It takes Hector 40% longer than Alex to do the same job. They decide to work

together. How long does it take them to complete the mowing when they work together?

Answer: ______________________________

2. Gayla and Christiana drive together from their hometown to camp. Gayla drives the highway and maintains a speed of 65 miles per hour for 3 hours. Christiana drives in town and maintains a speed of 25 miles per hour for 4.5 hours. What is the distance from their hometown to camp? What is the girls' average speed? Circle the column entry that contains the correct values.

Distance	292.50 miles	307.50 miles	307.50 miles	637.50
Average speed	41 mph	41 mph	45 mph	45 mph

3. Twelve art students sign up to paint the wall behind the catcher. The wall is one fifth of a circle. The radius of the circle is 14 meters. The wall is 5 meters high. The students paint one 22-square-meter section on Friday evening. About how many square meters remain to be painted?

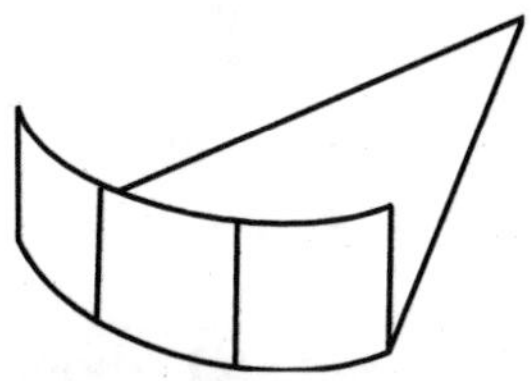

A 18 **B** 21 **C** 44 **D** 66

4. Ric takes underwater photographs at a depth of 16 feet below the surface. He signals to Steven when he starts his descent to 100 feet below the surface. Steven expects Ric to signal him when he gets to the new depth. Ric descends at a rate of 12 feet per minute. How many minutes does Steven wait before he gets a signal from Ric?

Answer: ______________

☆ *Challenge Problem*

You may want to talk this one over with a partner.

The community rallies to help build a sandbag wall to stop the channel silting in. You and 50 other people stand in line and hand sandbags from one to the other. On average, it takes five minutes to pass one sandbag down the line. Every half an hour, ten more people arrive and five people leave. In three hours, how many people are in the line? How long does it now take for a sandbag to reach the end of the line? Explain your answer.

____

____

____

Unit 7 Review

Review What You Learned

In this unit you have used mathematics to solve many problems. You have used mental math and estimation, practiced basic operations, solved equations, used statistics and probability, and both standard and metric measurements.

These two pages give you a chance to review the mathematics you used and check your skills.

✔ Check Your Skills

1. The school with the least participation per student population still manages to win the award for the most elegant robot design. Which school wins this award? Circle that school on the chart. Write an expression that gives the percent of its student body participating.

	Clifton High	Redlands High	East Heights High
Total school population	241	297	424
Program participants	16	25	29

Answer: ______________________________

If you need to review, return to lesson 1 (page 80).

2. Tobi recruits twice as many students for the project as Erica. Six of the recruits are the same. Which equation describes these facts? (Use x for Tobi's unique number. Use y for Erica's unique number.)

A $x + 6 = 2(y + 6)$ **C** $2(x + 6) = (y + 6)$

B $x - 6 = 2(y - 6)$ **D** $2(x - 6) = y - 6$

If you need to review, return to lesson 1 (page 80).

3. Austin approaches an intersection. A sign obscures the posted town names. Where is he?

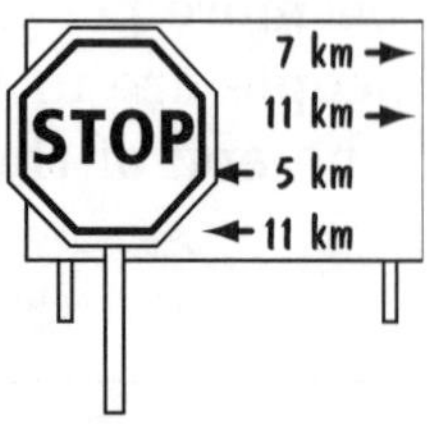

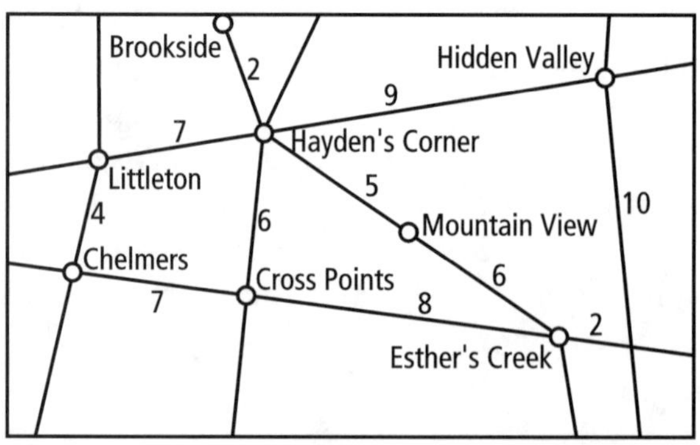

Answer: ______________________________

If you need to review, return to lesson 3 (page 84).

4. Out of fifty-four 11th grade students and forty-two 12th grade students, 30 sign up for volleyball. One sixth of the 11th grade signs up. What fraction of the 12th grade signs up?

Answer: ________

If you need to review, return to lesson 4 (page 86).

5. Marilyn takes 6 hours to exercise and train the horses. It takes Junesh 25% longer than Marilyn to do the same job. They decide to work together. How long does it take them to do the job when they work together?

Answer: Together they do the job in

______________________________.

If you need to review, return to lesson 5 (page 88).

Write Your Own Problem

Choose a problem you liked from this unit. Write a similar problem using a situation and related facts from your own life. With a partner, share and solve these problems together. Discuss the mathematics and compare the steps you used. If you need to, rewrite or correct the problems. Write your edited problem and the answer here.

Unit 8 Unemployment

Preview

How You Will Use This Unit

Unemployment is a time to think and plan. You may collect unemployment while looking for a new job. You may study the statistics and consider other jobs or careers. You may want to do networking, volunteering, or developing other skills. You may even think about consulting or contracting. As you consider your options, you will often use math. The math skills you use include mental math and estimation, basic operations and equations, statistics and probability, and measurements (standard and metric).

What You Will Do in This Unit

In this unit, math steps demonstrate how to solve problems. These steps can help you answer questions such as these:

The company that you work for cuts back its workforce. You lose your job. Your unemployment benefits include 40% of your salary. You earned $19,800 a year. What do you receive per week?

You are laidoff, and send out 24 job applications. Statistically, one in six applications results in an interview. How many interviews can you expect?

You register at a career tech program at a local education center. Of the 25 students, 9 are high-school students from local schools. What percent of students does this represent?

You average 50 kilometers/hour for half of the distance to your interim job. It takes you half an hour. For the rest of the way, you average 30 kilometers/hour. How long does the journey take?

What You Can Learn from This Unit

When you complete this unit, you will have used mathematics to work problems related to unemployment. These problems are similar to those that may actually occur in your daily life.

Lesson 1

Layoffs & Cutbacks

Example The company that Donova works for lays off 20% of its workforce. As part of this reduction, Donova loses her job, so she claims unemployment benefits. She learns that she can receive 40% of her salary weekly, not to exceed a maximum of $358. Donova currently earns $19,200 a year. What will she receive per week while unemployed?

Solve

Step 1: To find how much Donova receives in unemployment benefits, first, calculate her weekly salary.

$19,200 ÷ 52 = $369.23
There are 52 weeks in a year.

Step 2: Now, calculate 40% of this amount.

40% of $369.23 = 0.40 × $369.23
= $147.69

Step 3: Check to see whether this amount exceeds the maximum allowed.

$147.69 is less than the maximum of $358.

Answer the Question

Step 4: Donova will receive $147.69 per week.

✏ Now try these problems.

1. The company that Corey works for layoff 10% of its workforce. As part of this reduction, Corey loses his job, so he claims unemployment benefits. He learns that he can receive 40% of his salary, not to exceed a maximum of $358. Corey currently earns $38,800 a year. What will he receive per week while unemployed?

   Answer: Corey will receive ________ per week.

2. *MarketDays* layoff 3 out of 5 of 25 staff members. What is the workload per person after the cutback if the total workload stays the same?

   A same **B** $\frac{1}{10}$ **C** $\frac{3}{25}$ **D** $\frac{3}{5}$

3. Four friends lose their jobs and start a lawn mowing and trimming business. It takes all four of them four days to do four lawns.

   a. How long does it take one of them to do one lawn?

      Answer: ___________

   b. The desk calendar shows the date of the first day. Change the date to show the first date available for another job.

4. A local sports equipment company layoff its 15-person marketing department. The company saves $525,000 a *year* in salaries. The 15 people form a new company of their own. Very soon, they are selling their marketing services for an average of $5,000 per *month* per customer. Today they celebrate their first goal. The goal is to exceed the money that they made at the sports equipment company. How many customers do they have?

   Answer: ___________

☆ *Challenge Problem*

You may want to talk this one over with a partner.

A local computer manufacturer needs to layoff one third of its 300-employee workforce. The pie chart shows how these employees are divided into departments before the layoff. The Board of Directors recommends that the company cuts each department by one third.

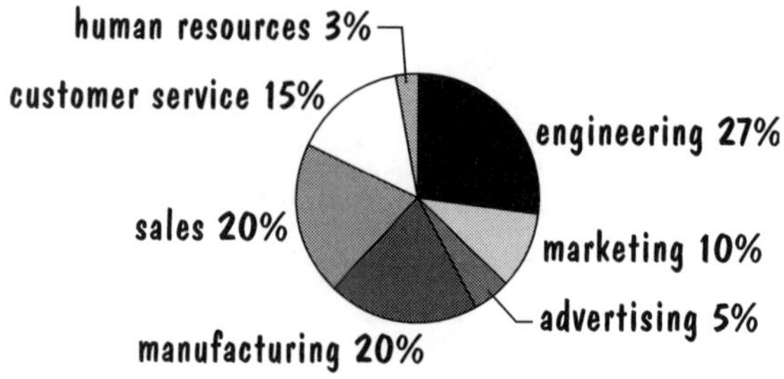

a. What is the impact on the smallest department?

____

b. What, in your opinion, are the positives and negatives of this strategy? What would you do instead? Explain your answer.

____

____

____

____

____

Lesson 2 Job Opportunities

Example Reza is laidoff, and sends out 24 job applications. Statistically, one in six applications results in an interview. How many interviews can Reza expect?

Solve

Step 1: Underline the sentence that tells you the chance Reza has of being called in for an interview.

Statistically, one in six applications results in an interview.

Step 2: Write the same sentence using the language of math.

The number of applications that result in an interview = $\frac{1}{6}$

Step 3: To find the number of interviews that Reza can expect, multiply by the number of applications sent out.

$\frac{1}{6} \times 24 = 4$

Answer the Question

Step 4: Reza can expect 4 interviews.

Now try these problems.

1. Jimena is laidoff and sends out 42 job applications. Statistically, one in six applications results in an interview. How many interviews can Jimena expect?

   A 1 **B** 6 **C** 7 **D** 36

2. Reed buys a phone card for $25 to make job application calls. Card calls cost 7 cents per minute within the area and 10 cents per minute outside the area. Reed makes two hours of calls within the area and one hour of calls outside the area. The rest

of his calls are outside the area. How many minutes does he have left on his card for outside calls?

☎ Phone card

$25

7 cents within area code
10 cents outside area code

Answer: ________ minutes

3. Abigail's telemarketing company needs to hire a new employee. The company is made up of herself, as manager, and six employees. Abigail creates an advertisement. She includes her own salary of $68,000 per year in her calculation of the average salary.

Great Job Opportunity!
If you enjoy talking to people, this job is for you...
Average Salary: $26,000 per year.
Call: 1 -2375

a. Which expression could you use to calculate the average salary of the six employees?
   A $\$(26 \times 7 - 68) \div 6$
   B $\$(68 + 6 \times 26) \div 7$
   C $\$(26 \times 6 - 68) \div 7$
   D $\$(68 + 26 \div 6) \times 7$

b. What is the average employee salary? What salary should be advertised? Why?

Answer: b. ____

4. Kacie goes to a local job fair. The fair lasts three days, eight hours a day. There are 84 one-hour information sessions that run, back-to-back, from 10 a.m. until 5 p.m. each day. (The last session each day starts at 4 p.m.) How many choices does Kacie have at any one time during the job fair?

Answer: ______ choices.

☆ *Challenge Problem*

You may want to talk this one over with a partner.

You buy office supplies, including letterhead paper, to use in applying for a new job. The bill in your pocket fades. What are the missing numbers?

_.8.27
33._7
5_.3_
Thank you

____

____

Lesson 3 — Changing Views

Example Giles takes one of three career tech programs at a regional technology center. His program has 12 high-school students from local schools, 6 high-school graduates who now seek a marketable skill, and 8 people who lost their jobs and want to change their career direction.

a. Which career tech program (A, B, or C) represents the program that Giles is taking?

b. Which career tech program data has an error? How do you know?

Career Tech Programs	Career Tech A	Career Tech B	Career Tech C
High-school students	46%	45%	46%
High-school graduates	23%	24%	23%
People who lost their jobs	31%	31%	33%

Solve

Step 1: Use mental math to add up the total number of students in Giles' program.

$12 + 6 + 8 = 12 + 8 + 6$
$= 26$

Rearrange (by commuting) the last two numbers so that the compatible numbers 12 and 8 are together.

Step 2: Now, find the percent of the total that each group of students represents.

High-school students $= \frac{12}{26} \times 100\% = 46\%$

Graduates seeking a marketable skill $= \frac{6}{26} \times 100\% = 23\%$

People changing career direction $= \frac{8}{26} \times 100\% = 31\%$

Step 3: Compare these percent values with each career tech program (A, B, or C).

This set of percent values matches Career Tech A.

Step 4: Now, add the percent values for each career tech program. Find the error.

$46\% + 23\% + 31\% = 100\%$	OK
$45\% + 24\% + 31\% = 100\%$	OK
$46\% + 23\% + 33\% = 102\%$	Error; the total should add to 100%.

Answer the Question

Step 5: Giles is taking career tech A. Career tech C has an error. The percent total should add to 100%.

✏ Now try these problems.

1. Robin registers for a home economics program at a local education center. It has 9 high-school students and also has 6 high-school graduates who now seek a marketable skill. It also has 14 people who lost their jobs and want to change their career direction. What percent of the program does each group of students represent?

Answer: High-school students: ______%. Graduates: ______%.

People changing career direction: ______%.

2. Ila-Mae loses her customer service job. Her pay was $120 per week. She likes to write, so she enters a writing competition. The entry fee is $20. The prize for a short piece of fiction is $2,000. She takes eight weeks to write a piece of short fiction, and wins. How does her weekly rate as a writer compare to her previous pay rate?

A $128 greater **C** $248 greater
B $130 less **D** $250 less

3. Todd volunteers for the coast guard and gets to ride in the coast-guard helicopter. The pilot tells him that they will fly at a height of 148.5 feet above the ground. He adds that the flying height times 8,000 gives the square of the visible distance. How many miles will they be able to see?

Answer: Visible distance = _______ miles

Hints:
1. There are 5,280 feet in one mile.
2. Height and distance must be in the same units.

4. Out of 40 people, 9 people typically come to the weekly networking meetings. On average, Ardean comes 4 out of every 10 weeks. What is the probability that she'll be there next week?

Answer: __________

☆ *Challenge Problem*

You may want to talk this one over with a partner.

You run around the rectangular school sports field three times. The area of the field is 2,450 square meters. The length of one side is twice that of the other. What are the dimensions of the sides? How far do you run? Explain your answers.

____

____

Lesson 4 Putting the Pieces Back Together

A Puzzle Game (for Five Players)

The goal of this game is to make squares with three of the fifteen puzzle pieces. First, you try to fit your three pieces together. Then each player passes one piece to the next player until all players have made a square.

Materials

Puzzle pieces (on the next page), mixed up and separated into five sets of non-matching threes.

Directions

1. Give a set of puzzle pieces to each player. Sit with players around the playing area.
2. Each player examines the pieces to see if they fit together to make a square. Each player then passes (discards) one non-matching piece to the player on the left.
3. As the game progresses, players examine their own pieces and watch those of the other four players to see what pieces are needed. No player can say anything. (The game is played in silence.)
4. Players repeat the process until all players have made a square.
5. The game can be expanded for more players by adding more sets of three pieces that make squares.

Before you play the game, try these warm-up problems.

1. You and some friends play *Putting the Pieces Back Together.* You have the only piece that has more than six sides. What other pieces do you need to complete a square using this piece?

   Answer: ______________________________

2. You all examine the pieces for *Putting the Pieces Back Together.* How many four-sided shapes are there (excluding the four large, complete squares)? How many similar triangles are there?

   A Two, five　　**C** Two, eight

   B One, eight　　**D** Four, six

Puzzle Pieces for *Putting the Pieces Back Together*

These diagrams show five same-size squares made up of sets of three pieces each. Copy (and enlarge if you want to) and cut the pieces, mix them up, and separate them into five sets of unmatching threes.

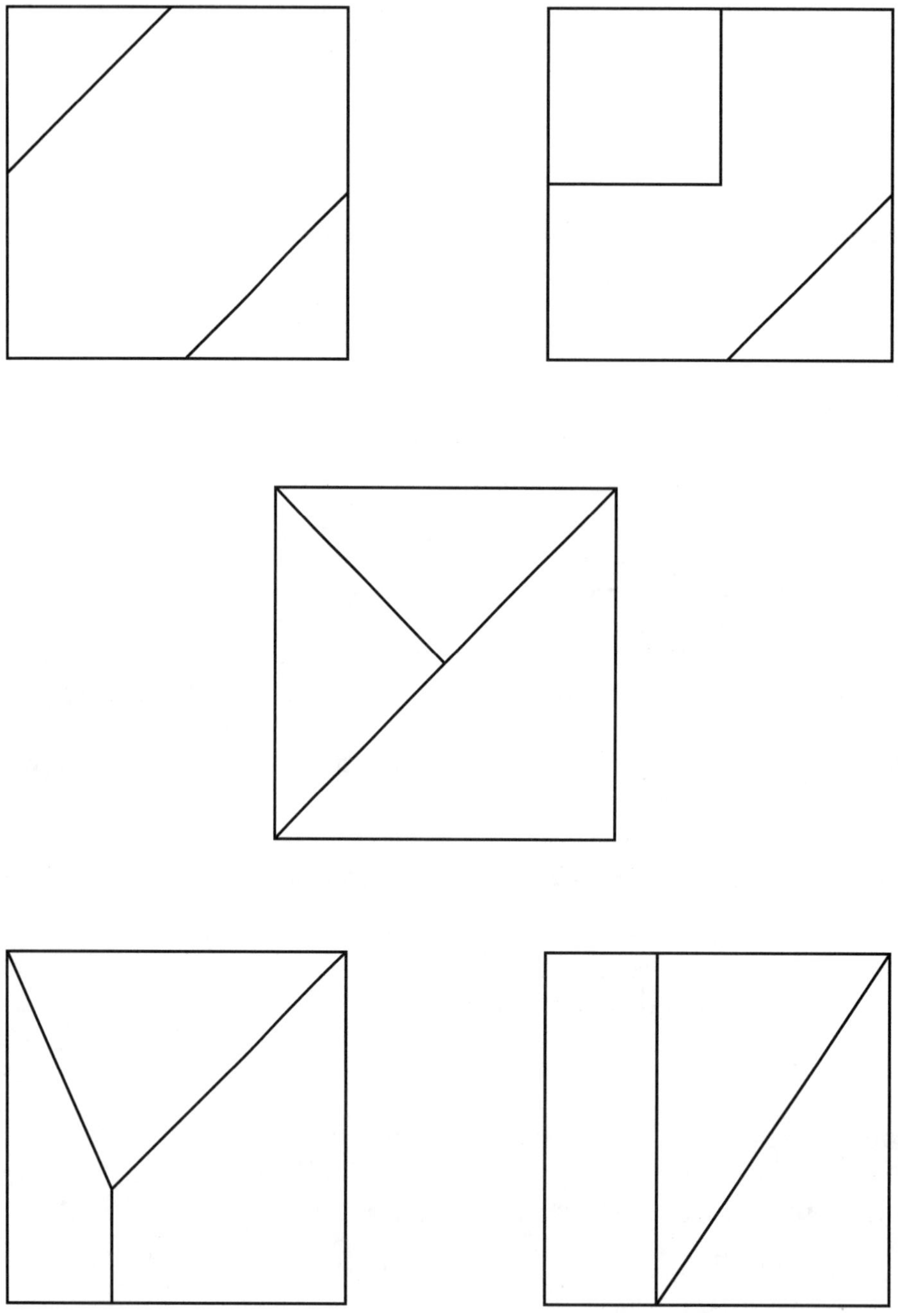

Lesson 5 — *Interim Jobs*

Example Barry averages 80 kilometers per hour for $\frac{2}{3}$ of the distance to his temporary job. This part of the journey takes him one hour. What is the total distance that he travels? For the rest of the journey, he only averages 60 kilometers per hour. How long does the total journey take him?

Solve

Step 1: Use d for the total distance that Barry travels. Write an expression for the distance he travels on the first part of the journey. Solve for d.

$\frac{2}{3} \times d = 80 \times 1$ — Distance = speed (80 kilometers per hour) × time (one hour).

$d = 80 \times \frac{3}{2}$ — Multiply both sides by $\frac{3}{2}$.

$d = 120$ kilometers — The total distance is 120 km.

Step 2: Now, find the distance Barry travels on the second part of the journey.

Distance on the second part $= \frac{1}{3} \times 120$ kilometers
$= 40$ kilometers

(The first part of the journey is $\frac{2}{3}$ of the total distance. So, the second part is $\frac{1}{3}$.)

Step 3: Now, use t for the time that the second part of the journey takes. Use information in the third sentence to write an expression for time, t.

$40 = 60 \times t$ — Distance = speed (60 kilometers per hour) × time.

$\frac{2}{3} = t$ — The time is $\frac{2}{3}$ of an hour.

Step 4: To find the total time that the journey takes, add the time for each part.

Total time for the journey $= (1 + \frac{2}{3})$ hours

Answer the Question

Step 5: The total distance is 120 kilometers. The total journey takes Barry $1\frac{2}{3}$ hours.

Now try these problems.

1. Carli averages 35 kilometers per hour for half of the distance to her interim job. This part of the journey takes her three quarters

of an hour. For the rest of the journey, she averages 60 kilometers per hour. What is the total distance that she travels? How long does the total journey take her?

Answer: Carli travels ________ kilometers.

The journey takes her ________ hours.

2. Between full-time jobs, Tocha writes short articles for local papers. During November, he writes 15 articles. He earns $1,200 each for three articles, $850 each for four others, and $400 per article for the rest. What is his mean pay per article?

   A $493.33 **B** $680.00 **C** $816.67 **D** $925.00

3. To gain experience, Hayes works on commission for a hardware store. One week he sells $2,050 worth of products and gets $61.50 commission. The next week he earns $59.43 in commissions. In which week does he earn more money? Fill in the blanks. Circle the week.

Week 1	
Sales:	$2,050.00
Commission:	$ 61.50
Total:	________

Week 2	
Sales:	________
Commission:	$ 59.43
Total:	________

4. Beki takes two part-time jobs, one following the other. At one job, she earns $9 per hour for six forty-hour weeks in a row. At the second job, she earns $1,600 for a four-week month. What is her mean pay per week over this ten week period?

   Answer: ________

☆ *Challenge Problem*

You may want to talk this one over with a partner.

You volunteer on two task forces, along with 23 other people. One third of task force A members are on task force B. One quarter of task force B members are on task force A. You are on both task forces. How many *other* people are on both task forces? How many people are *only* on task force A or task force B? Draw a diagram to explain your answer.

____

____

Unit 8 Review

Review What You Learned

In this unit you have used mathematics to solve many problems. You have used mental math and estimation, practiced basic operations, solved equations, used statistics and probability, and used measurements on figures of different shapes using both standard and metric measurements.

These two pages give you a chance to review the mathematics you used and check your skills.

✔ Check Your Skills

1. The company that Marty works for cuts back 25% of its workforce. As part of this, Marty loses his job. So he claims unemployment benefits. He learns that he will receive 42% of his salary. Marty currently earns $24,600 a year. What will he receive per week while unemployed?

   Answer: ___________

   If you need to review, return to lesson 1 (page 93).

2. Zia is laidoff, and sends out 18 job applications. Statistically, one in six applications results in an interview. How many interviews can Zia expect? Shade the number of applications that result in interviews.

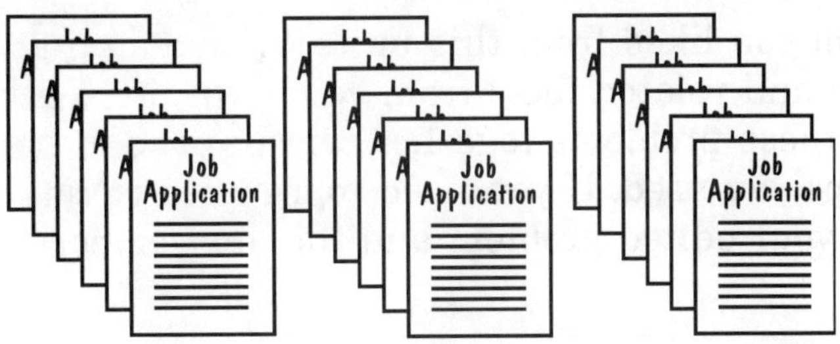

   If you need to review, return to lesson 2 (page 95).

3. Terry registers for a career tech class at a local education center. She is one of 6 high-school students from local schools. The high-school students make up one fifth of the class. The class also has 12 high-school graduates who now seek a marketable skill. The remaining students lost their jobs, and want to change their career direction. Which statement is *not* correct?

   A High-school students: 20%
   B High-school graduates: 40%
   C Remaining students: 40%
   D Number of remaining students: 15

   If you need to review, return to lesson 3 (page 97).

4. Out of the 23 people at the networking meeting, 12 work in industry, 6 people teach, 3 work for the Tourist Bureau, and 2 work for the city. What percent does each group represent? Match the percent to the group.

Group	Percent of people at meeting
Industry	52%
Teaching	9%
Tourist Bureau	26%
City	13%

If you need to review, return to lesson 3 (page 97).

5. Andri averages 50 kilometers per hour for half of the distance to her summer job. This part of the journey takes her half an hour. For the rest of the journey, she averages 30 kilometers per hour. What is the total distance that she travels? How long does the total journey take her?

Answer: The total distance is _______ kilometers.

The journey takes her _______ hours.

If you need to review, return to lesson 5 (page 101).

Write Your Own Problem ✍

Choose a problem you liked from this unit. Write a similar problem using a situation and related facts from your own life. With a partner, share and solve these problems together. Discuss the mathematics and compare the steps you used. If you need to, rewrite or correct the problems. Write your edited problem and the answer here.

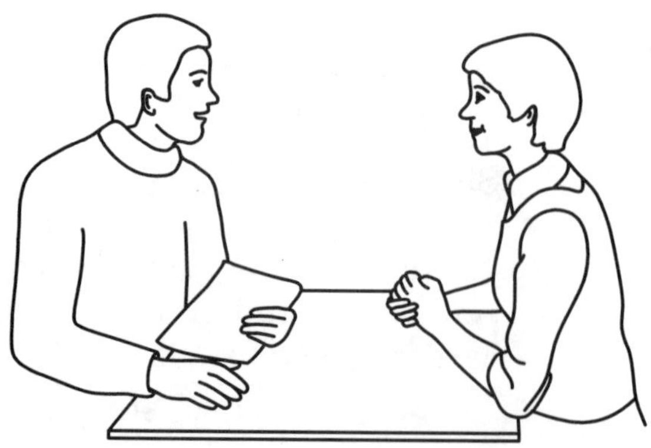